My Journey Into Myth and Mystery

The Search for Sasquatch

By

K.J. Blount

Things are not always what they seem; the first appearance deceives many; the intelligence of a few perceives what has been carefully hidden"

Phaedrus

Dedication

To my son Francisco. For always having my back when I need you...

COPYRIGHT PAGE

ISBN-13: 978-1512083590

MY JOURNEY INTO MYTH AND MYSTERY
THE SEARCH FOR SASQUATCH

COPYRIGHT 2015 K.J. BLOUNT
ALL RIGHTS RESERVED

INTERNATIONAL COPYRIGHT PROTECTION IS RESERVED UNDER UNIVERSAL COPYRIGHT CONVENTION AND BILATERAL COPYRIGHT RELATIONS OF THE USA. ALL RIGHTS RESERVED, WHICH INCLUDES THE RIGHT TO REPRODUCE THIS BOOK OR ANY PORTIONS THEREOF IN ANY FORM WHATSOEVER EXCEPT AS PROVIDED BY RELEVANT COPYRIGHT LAWS.

PUBLISHED BY RS PUBLISHING AND DISTRIBUTION
HTTP://WWW.GHOSTSOFRUBYRIDGE.COM/RS-PUBLISHING-AND-DISTRIBUTION/
PRINTED BY CREATESPACE INC. **WWW.CREATESPACE.COM**

PRINTED IN USA

ISBN-13: 978-1512083590

EDITED BY: THOMAS D. "THOM" CANTRALL

COVER: PHOTO BY BRIAN BLAND IN THE LAND OF ARRRESHNAH'S CLAN, BRITISH COLUMBIA, CANADA

Table of Contents

Timeline of Events Covered in this Volume

2008 Hand print and bed find in sand dunes

2008 Cemetery area - bed, gift, foot print in mud, walk through field, sighting, flash of light, took Cristy

2008 Drove to hunting cabin found hand print on shore and finger marks on hill

2008 Found trackway in beach grass leading from trees to shoreline and back at spillway

2009 Found lake, cast footprint, Frank finds second print

2009 Cristy and I found track way up creek

2009 Patrol on hill, fed sasquatch hot dogs

2009 At lake alone, protected

2009 Orb across lake

2010 Moved to Oregon

2010 At lake with Friend

2010 Went with Cristy and Friend again for divorce weekend. Sneaked up on them tree by tree

2010 Went to OSS in Eugene

2010 Had contact in room. Hands on head. Hair floating, hand on back of head. Cristy called said something opened my bedroom door and walked down hall. Friend was touched.

Scent in bedroom changed from scat to orange blossom smell

2010 Woke up and my feet looked very unfamiliar to me and my legs looked very puny and small.

2010 Went up Memaloose

2010 Returned to Memaloose

2011 Moved to Winlock, WA

2011 Went to OSS with Thom outside of Sisters, OR. Played flute on tree across river, had toiletries hidden in morning, Charles fell asleep and they walked into camp.

2011 Went with lady for summer, met with other researchers, back to lake, over to Cristy's for surgery, moved in with Frankie in October

Cristy moved to Coos Bay

2012 Primal People conference in Washington, camped in Blue Mountains at cabin. Found footprint in snow, played flute and Cristy had sighting that looked like it dissolved into a bee swarm, took lady to amphitheater to describe sighting-later at cabin something stood in front of Cristy. Camped at Hobo camp for a week and had something hit the truck and heard things on the roof.

2012 Camped at lake with Connor for first time. Peed first night behind Cristy's tent, milk stain on blanket, Given towels with one rock, one sat near us, one walked up to camp, three walking around in camp, found baby print. Friend and son came and found a large track and elk tracks, a den area for hunting, and an arch tucked under a log. Friend and I watched an orb float in out of the trees and hang there in front of us before flying off to our right and into forest. Friend told story of how something left an extra propane tank lid that actually fit their propane tank after talking about how they were missing one.

2012 Camped at Powers for first time. Cowboy and I heard sounds above the little hillside, next night one was watching from tree copse, ran to outhouse and threw rocks into camp at me. Big one came across the river and stood behind large boulder. Found prints in camp and sand next morning, destroyed them.

2013 Friend and I find deserted Campground, she gets poked, wand left in mud where I peed.

2013 Cristy and Connor and Caydence camping at skull rock with other friend, found glyphs in front of tents.

2014 G-O road campout

2014 Camping with Friend at Skull rock. Saw shimmer in forest, heard wailing above us, had two in camp, found footprint in gravel

Foreword
by
Thom Cantrall

I have been around this field of research for many, many years. In fact, those who predate me are few in number. The one thing that has kept me going and interested all these years is the quality of friends I've made. Over the years, there have been many... most did not stay the course and went their own way. Some, like Kathi, have remained true to their calling and have learned and have grown in this quest. It has been my pleasure to have her and her friend Cristy in my camp in three different states and have found them to be steady, worthwhile mainstays in any kind of camp environment... and always up for providing a laugh or two, even if at their own expense.

It has been my pleasure to have been a part of many of the experiences mentioned in this volume and I can attest to the truth of that which she reports. The night at Memaloose was precisely as reported... the days at the Toll Gate cabins as well. The most precious were the fourteen days we spent in June of 2014 in the Siskyou Mountains of far northern California. It was a most magical place with a most magical cast of people. I do believe everyone who stayed in that camp for the entire time had an encounter.

Most memorable for me occurred on our way back into camp from town. I had two people in the car with me, Arla in the back seat and Jackie from England was riding shotgun in my Buick. As we rounded a curve near mile post fourteen on the Gasquet-Orleans road, immediately in front of us were two very large, very tall sasquatch! At a range of some twenty to twenty five feet we watched two very real "myths" exit the road in record time!

Next most important, to me, were the events in the story Kathi related in Chapter Twenty of this volume. She told the story well, so I will leave that to her. That time spent in that beautiful place was so intense that people who were there have asked to return this summer.

Read Kathi's work here with an open mind. It is important to understand that there are so many wondrous things that happen to us that this book of wonders is but one more step in understanding our lives. Kathi has moved out from her comfort zone to share with us the events of the past seven years that have molded a new understanding for her.

Use her information in your own quest for, I assure you, it is totally true as told.

Thom Cantrall

Prologue

My fondest childhood memories are of camping trips taken with my family and of sunny days spent in the mountains by a river or lake. Growing up, I was told exciting stories of my ancestor's wandering the wilderness of the new west, mapping the terrain and photographing nature. My uncle worked for the Forest Service for as long as I could remember but not once was the subject of bigfoot ever mentioned. No stories were ever told of hearing strange sounds in the night or seeing anything beyond the ordinary. I never even saw the film footage of the bigfoot crossing a sand bar in California until 2008, and I grew up in California!

I spent a great deal of my time alone roaming the mountains and foothills in California when I was a child. I married a logger and spent a lot of time in the mountains in the various places he worked. We had a few strange experiences while we were out there, but it never occurred to us that it would be anything other than a natural occurrence done by the usual wildlife that lived in the area.

I remember one day specifically when my two dogs were walking with me... Suddenly they both did an about face and began dragging me backwards. One dog had my arm, the other had my pants leg and they were physically dragging me straight through the bushes and trees in a bee line for the truck. These were large dogs and not afraid of wildlife. I never could understand why they had done this.

All that changed for me when we moved to Washington state. I had no prior knowledge of this species of life and it provided me quite a surprise to find out that they do indeed exist... They are a real species, and they really do live here right alongside us... And they live closer to us than I could have ever believed.

Since this journey began for me in 2008 I have recorded my experiences in my journal as they happened. I have had some amazing adventures in the woods and have been told that I needed to share these experiences with others as maybe it would prove useful to someone who has also had to deal with the same reality slams as I have. Maybe it would just be good information for others to have about this species of life.

I have started to write this book time and time again. I would start and then stop, start and then stop, over and over again. Each time I thought it would be safe... I thought that maybe other people were finally ready to hear stories like mine, I would chicken out and set it back. I knew that many people were going to say that I "must be making this information up" and that still others were going to say that "it's just plain impossible for some of these things to have occurred." I knew that by opening myself up about the things that I've seen, I would also be opening myself up to more criticism and controversy than I ever really wanted to deal with in my lifetime.

I also know that it's hard for people who haven't had these experiences to even believe in the possibility of them. Hey, I know how you feel, I was once like you! Only after seeing these strange things, and after living through these strange occurrences was I forced to become a believer. After all, it's hard to continue doubting what you've already seen and done. I just ask that the reader keep an open mind and consider the possibilities, and if you cannot do this, just enjoy it as a good story.

I truly think that this is the time for this story to come to light. Others before me have already stuck their heads out and admitted that these things have happened to them. Does this make me a coward for waiting so long? Perhaps... but perhaps not. I do know that the world deserves to hear these stories now, right alongside the more mundane and safe sighting reports.

I think those of us who have had the "impossible" happen to us, or right in front of our eyes, deserve to be taken seriously. If we are ever going to get to the bottom of what this species really is, then we need to know about every aspect of every encounter, not just the ones that make us feel safe. We need to know exactly what they are capable of, and then find out whether each and every sasquatch has the same level of capabilities.

I don't know why some people get to have these mind changing experiences and yet others can spend a whole lifetime in the woods and see nothing. I wish I could tell those who ask. We could then all share in these magical sightings and "off the wall" experiences. I have learned by experience that some people will never have a positive encounter. I've witnessed this by taking people with me that completely stopped any and all contact simply by being in the vicinity. I didn't feel that they were any different from me but the sasquatch obviously did and they showed it by either staying away completely or by marking these people with a horrible scent.

It seems that everyone who does have a sighting or does have these experiences has a change of heart and a change of beliefs, both emotionally and spiritually. I was forced to believe in something that my mind had never even contemplated before. I wasn't thinking about sasquatch when I began this journey. I sure as heck wasn't out looking for orbs when I was forced to admit that they exist by seeing them with my own two eyes. These experiences force you to admit that there is more to life than only what you have seen, or believe in. And there is a lot more out there than what we were ever taught in science class.

I also began to look at how we are treating the earth and all creatures that live on it. I began to think of the effect we are having on our oceans and all who live in them. In other words, the profound shock of knowing that there is more out there than I ever knew about woke me up…completely! I know that this species could wipe out all of human kind if it truly wanted to, but they don't seem to want that at all. They have always been kind to me and the love that I feel while they are around is completely mind altering.

While growing up I was taught that man is the top of the food chain. That we control all other creatures that live, and that we are the most important things alive, but that couldn't be farther from the truth. We are not the top of the food chain, and we are not here to control other species but to protect other species.

I know that I am speaking the truth about what has happened to me, and I know that there are some very strange things happening in the woods. I hope all can enjoy what is presented here and I invite all to come with me on my journey.

Kathi Blount

Chapter 1 – The Journey Begins

My journey began on April Eleventh, 2008. I remember that day so well that I don't think I will ever be able to forget it. Until this time the name bigfoot or sasquatch, meant nothing to me other than being the subject of a great fictional movie, "Harry and the Henderson's." I never once gave thought to the fact that it could have been based on actual events, or that people were actually seeking this creature in real life. I came from a family who loved the outdoors and I have spent a great deal of my life alone out in the woods and mountains yet I had never seen or heard of this species, until…

My husband and I were living in our motor home in a very small coastal town in Washington. It had been raining for five days continually and being stuck in the motor home was not fun. I was suffering from "cabin fever" from having stayed indoors so much. As we were drenched as soon as we stepped foot outside, hiking had not been an option.

When I opened my eyes that morning, the sun was shining brightly and the air was nice and warm. I jumped out of bed and grabbed a cup of coffee, and started stuffing the things I would need for the day into my bag. I was headed OUTSIDE, and the housework could wait. I needed the sun and the fresh air, and I was going alone this time.

I owned five cats that we had rescued from various situations and they liked to tag along with me while I went walking through the woods near the motor home. They liked to chase each other through the trees and ferns while I walked along, but I had noticed a pack of coyotes hanging around over the past winter and I feared for my kitties' safety, so I locked the cats in the house and made them stay behind.

I had a favorite spot that I liked to go to with the cats where I could just sit and read. It was a beautiful grassy area with tall trees and berry bushes growing all around In all the time I had been coming here I had never seen another human being, which made it the perfect spot for me. I'm more a loner than the life of the party, so I usually seek out places where I can be away from others. When I reached this spot on this day however, I decided that since I didn't have the cats following me I would hike in further and see what other scenic areas I could find here.

Figure 1 One of my kitties following me through the woods

I walked through the trees until they opened onto a large swampy area filled with beach grasses that grew about knee high on an average person. I stopped to look at it because the grasses looked so lovely blowing in the wind. The water had already receded in most of the field and as I looked, I noticed that a relatively new trail had been forged across the grasses. I wondered about that and went over to check it out. It didn't appear to have been made by a small animal or even a bear or coyote. Both of these animals were known to be in the area, and I had no desire to run into either one of them. The grasses weren't brushed over along the tops like they would be if a shorter, four legged critter had trotted through the grass. It was squashed in places like I would have done if I had walked through the tall grass... So I wondered who had been out here during the storm and where they were going. I decided to follow the trail and see where I came out.

I followed the tracks until they led to a sandy beach access trail. I was thinking at that point that someone else must have wandered in and found my secluded spot and that I was now going to have to find another place to hang out. I turned to the left and followed this beach access trail for a short ways until it led me out of the sand dunes and onto the beach. I didn't really feel like going to the beach that day, so I backtracked to the original trail. I was standing there surveying the scenery when I noticed a small game trail leading into the woods across from where I now stood. I decided to follow this small path and see what I could find.

As I rounded a bend in the trail I saw that a very large tree had fallen over and was completely blocking my way. I literally had to climb up into the tree, and then through its branches to get to the other side. When I did get across the tree, I found myself in a beautiful clearing that was totally encircled with large evergreen trees. There were two sand mounds to my right which were covered in sharp, pointed beach grass and had a trail running between them that continued on into another field of grass and evergreens. There were two wild apple trees growing off to my left and a small hillock in front of me.

Since it was totally surrounded on all sides by trees, and the dead fall that lay on the path behind me, it was a perfectly secluded spot. I couldn't even see the trail that I had just walked. My first thought was what an awe inspiring spot this was. My second thought was that I wanted to make this my new area.

I set off adventuring. I followed the game trail towards the hillock in front of me and noticed a nice shady area under a small tree at the bottom of it. I headed for that tree to sit down and eat my lunch. That's when I noticed that I hadn't seen the two crows that normally followed me who I nicknamed Heckle and Jeckle, after the two magpies of cartoon fame. Nor had I seen any birds for that matter, all morning long. That was very strange for this area. We had hundreds of birds that would normally be singing and flying through these trees every day. I always looked forward to the two crows coming in and sharing my lunch with me.

I was thinking about this when I arrived at the designated tree and began to unload my gear. I had put my bag down on the ground and was reaching inside of it to get my water bottle when it suddenly felt like someone was standing really close to me and staring at me... and not in a friendly fashion. The hair stood up on the back of my arms and neck and I felt extremely uncomfortable. This feeling was so strong that, instead of sitting down and eating I just stood there circling and looking all around me. When I first turned around I was completely expecting to see someone standing behind me, but I was entirely alone as far as I could see. I tried to tell myself that I was just being paranoid or something, but I couldn't shake the feeling that I was being scrutinized, analyzed, and stared at. The longer I stood there the more foolish I felt, but it was very distracting to me.

I started to look harder for anything that would explain this strong feeling of being scrutinized so deeply, but the only place that I could see where someone could possibly be watching me was a dormer window on a house that was showing through the trees. It looked to be about three or four blocks away and at the edge of the sand dunes. All I could really see through the trees was the upper corner of an attic window, but I told myself that this had to be what was bothering me. I didn't see how anyone could be standing there and staring at me from this window, or see how it could possibly explain the strong feeling of someone standing next to me and watching me so closely, but it was the only thing my mind could come up with at the time.

Since I was so uncomfortable in this spot I figured I'd just walk up and over the hillock and sit down on the other side so the house couldn't see me anymore. I know it sounds strange now, but it was the easiest explanation that I could come up with at the time and it was just easier than trying to deal with it.

I don't think my mind really wanted to deal with anything more at that point. I had just recently lost my mother and my head was abuzz with thoughts of her and what I was going to do without her. I didn't want, or need, any more stress or drama in my life at that time and it was just easier to move to a different spot than to try to figure out what was going on.

I walked over the hillock through that long, sharp beach grass, and when I got to the back side I spotted a game trail that wound down the hill and then passed between two medium sized trees. As I was approaching these trees I noticed a large round bark ball, or gall, growing on the tree to the left. I always loved those bark balls... Probably why I noticed it. My intention was to walk right between these two trees and continue forward on my way down the trail but for some reason my body stopped right before passing between them. I just stopped! My upper body actually rocked forward a little when my lower body stopped in its tracks.

At the same time that my body decided to stop, I got a message from my mind saying, "That is a bee's nest. It will sting you." Worded just like that.

WHAT? I said in my head. I am now going to replay that conversation word for word, just as it happened. I will use capital letters to show where emphasis was placed. I still don't understand how I had this conversation, or with whom I had this conversation, but I know that I did have this conversation in my head with a second party! It wasn't even my voice that I heard talking to me. Talk about life in the Twilight Zone! So, to sum this up... I am standing in front of two trees with a trail leading between them, in broad daylight, and my body refuses to move, and I'm talking to something else in my own head. Well, to be more precise, I am arguing with something else in my own head... Here we go.....

"That is a bee's nest. It will sting you."

No it isn't, it's a bark ball.

"It is a bee's nest. It will sting you."

It isn't A BEE'S NEST, and it's not going to sting me.

"Do not come here. It will sting you."

I WANT to come in there, and I'm NOT going to get STUNG!

"It Is a bee's nest. It will sting you."

It's NOT a bee's nest!

As I got to the words "bees nest" in that last statement of mine, I noticed that I had already turned around, and my body was actually starting back in the direction from which I had just come. I thought to myself, What in the heck am I doing? I wanted to follow that trail and go in there and find some shade, and.... Why am I walking? Whatever!

All this was just too much, and I truly think my mind just didn't want to deal with it because I just kept walking and went back the way I had come. I didn't feel scared that this conversation had happened; in fact, I remember just feeling numb at the time. It was starting to get hot out there that day and I just wanted to sit down and think about what the heck had just happened to me, so I started to look for a place where I could sit down that wasn't covered in that sharp beach grass.

I was having a bit of trouble dealing with the fact that I was hearing voices in my head. Was I going insane now? I mean, a lot of people lose their mom, but they didn't crack up and start having conversations with things that don't exist, about things that don't exist. Did they? No, I didn't think so.

As I approached the tree I had climbed over to get into this area, I looked off to my left and again noticed the trail leading through the two sand mounds that I have mentioned earlier. As I walked up to them, I noticed that the grasses on the one to my right were smoothed down on top of it. The grasses had been laid flat and pretty much in one direction, while the one on the left still had all the grass standing straight up on it, with the exception of one area which had a deep hole dug up out of it.

I didn't think much about that hole at the time. I'm sure if I did I probably attributed it to the coyotes. I also noticed that there was sand showing on the trail between these two mounds, and the only vegetation growing here was new grass shoots coming up through the sand. No moss grew here either, whereas the rest of the area was covered in moss and this beach grass. I didn't think about this fact at the time either, and I decided to sit down on the sand mound with the flattened grasses since they wouldn't cut me there.

As I began to sit down on the edge of this mound, I turned slightly to my right so I could put my bag down in the sand. As I did this, I was scanning the ground in search of a snake, or a snake trail. I am deathly afraid of snakes and try to avoid them at all costs. I saw the imprint of a long line in the sand, which I took to be a snake trail, so instead of dropping my bag in the sand, I turned and placed it on the mound behind me. I then started following this line with my eyes to see where the snake had gone, and I noticed that the trail ended in a berm of sand, instead of continuing on into the vegetation somewhere. It wasn't what I was expecting to see at all, and I remember saying out loud, "Where did the snake go?"

While I sat there trying to figure out how a snake could make a trail going forward, and then back up in the exact same trail, my back started screaming in pain at me. I had been leaning over to the right with my right hand resting on my right thigh, and as I pushed myself into an upright position my eyes were able to take in the whole scene on the ground in front of me. I became aware of more lines in the sand. One of them actually appeared to be a thumb print, but my brain was having none of that. It wasn't until I looked down at my hand on my thigh that my brain had to admit the obvious. I was looking at a huge hand print in the sand, and it was positioned exactly the same as my hand was positioned on my thigh. The pinkie finger and ring fingers were flat, the middle finger and index fingers were both bent and the thumb laid flat.

Something very large had to have pushed itself up in this sand! I put my hand into the imprint to see if I could be right and it was an exact fit, although the one in the sand was much larger than mine. I tried to tell myself that it was just a large man's hand print and nothing special, but this one was huge! As a woman I had to know if it was possible that a man, with hands that large, was hanging around with me out here alone. I decided that I had to show this to my husband.

When my husband came home, I took him out to this spot to show him the hand print. It was even larger than his hand, and he suggested we look around to see if we could find any more prints in the area. We never found anything else but, as I said, the ground here was covered in thick moss and beach grass except for that one sandy path between the two sand mounds.

He pointed out the fact that of the two mounds, only one was flattened completely. We agreed that we didn't know what could have caused that. We also noticed that the hole on top of the other sand mound had to have been dug by hand because we didn't see any claw marks and something was pulling the dirt up from the hole and laying it to one side, not spraying it out behind the hole like a dog would do.

I asked him what he thought had made the print, and what could have been out here digging holes in the sand, and he brought up the subject of a sasquatch. That explanation hadn't gone through my mind at all. I was thinking a very large man when I had originally seen the imprint in the ground and that was why I had thought to bring my husband out here. I wasn't too surprised that the imprint had been so much larger than my own hand when I had compared it because I have pretty small hands. I was surprised when I saw that it was so much larger than his so when he said it was a sasquatch and I couldn't come up with a better idea, I went along with it.

He suggested we cast the hand print to preserve it for later identification and said he knew just the product to do it with. He said we would use a product called Lockite and the casting would be as hard as stone and wouldn't break easily. We gathered the needed products and went back to where the imprint was located. It was still there and we mixed up the material and poured it over the print.

Figure 2 Area where handprint cast

As my husband already had plans to be away from the house, we had initially decided to just let the casting cure overnight and retrieve it in the morning. About an hour before dark, I watched the weather forecast and it called for rain for the next few days. I wanted to get that casting out of the ground before it was ruined and I was so excited about it that I decided to just go back out and get it myself. What could go wrong? As long as I was out of the area before dark, the coyotes probably wouldn't be a threat.

To save time I drove as close as I could get to the area in my truck then hiked the rest of the way to the site. I quickly located the casting and began to dig it out of the ground. I was using a camping spatula with a serrated edge on one side, as I didn't own a shovel. It was the largest thing I could find at the time to take with me, and I'm glad that I grabbed it. The casting material had run off the trail and down into the grasses that grew on the trail. That had essentially glued the cast to the ground. I had to use the knife edge of the spatula to cut the casting away from the plants. I then planned to use the spatula to lift the casting up and out of the ground. I had it all planned and I was having the time of my life.

I couldn't believe that there was even a glimmer of truth that these unique beings actually existed. I mean really… a real life Harry Henderson in my woods? And who would have thought that after all the time I had spent out in the woods alone, I would find something like this practically in my own back yard! I was in seventh heaven. I already had a place picked out on my wall where this unique gift was going to hang. Everything was going great, until I was about half way done, and then it hit… back to the Twilight Zone!

It began with emotions, and feelings that were not my own overtaking me. Remember, I was happy and having the time of my life! Suddenly a feeling came over me that was of total confusion blended with curiosity, like a "Hmmm…What are you doing there" kind of feeling. I froze for a second wondering where that came from but I shook it off and continued with my job. This feeling was followed closely by an intense confusion mixed with an overdose of concern. More of a "Hmmmm…Yeah, you shouldn't be doing that" kind of feeling. I brushed it off again and began to lift the casting up out of the ground. This would have been a lot easier if we had taken the time to look up the correct way to prepare a sandy soil for casting. Maybe that way it wouldn't have glued itself to the surrounding vegetation so securely.

The feelings I was getting weren't as bad as the voice that boomed into my head just then and asked, "Why are you taking that?" Followed by, "I cannot let you take that!" This was so intense that I instantly lost all my happy feelings and I got scared. I began shaking and I couldn't understand where these feelings were suddenly coming from. None of the things I had been through today could possibly be happening, and yet, they were happening. And I was out there alone, again! I started to talk out loud... well blabber out loud to be honest, and I said, "I just wanted to have it for me, to let me know that you're real."

At this time, I had literally torn the casting from the ground and was trying unsuccessfully to stuff it into my bag which was fighting me all the way. At some point during this exchange, I had gotten an intense feeling that he was hiding under this huge tree which was right behind me.

This tree had very lush branches all the way down to the ground and was impossible to see into, even in broad daylight. And it was anything but full light right at the moment. As soon as that thought became clear to me, I turned my body toward the tree so that this thing couldn't come up behind me. I kept waiting for something huge to lunge at me and rip this casting right out of my hands. Now the feelings and the words were becoming more aggressive, concerned and agitated all at the same time.

"I cannot let you take that; it is a piece of me!"

I cried out, "I swear I just want it for myself, I won't tell anyone else where you are, and I won't come here anymore."

At this point I had successfully stuffed the casting into the bag, ripping a nice hole along the seam of the bag in the process. I said, "This isn't me taking a piece of you, and I promise no one will ever know you were here!"

Then I heard something in my head that made me freeze where I was for a second or two and I had to think about whether what I was doing was a good thing, or a bad thing. I heard it say something about the fact that to take a piece of it was to take a piece of its soul. I hadn't been intending to take a piece of it, or to control it by having a piece of it, but was I doing a good thing for it?

I started backing up the game trail away from the two hillocks and towards the dead fall that led out of this area. During this time I could still feel the emotions of this creature, and I had tears streaming down my face. It was so concerned about me taking this casting, and so anxious and agitated all at the same time. It was panicking me. I was continually talking out loud and it was speaking in my head. Talk about utter chaos in my mind. When I reached the main trail I had to turn around to climb over the dead fall that was blocking my way, and I heard something step out from under the tree and crunch through the grasses as it followed me. At this time I lost my composure completely.

I was bawling like a baby and could barely see where I was going. I was walking as rapidly as I could without breaking into a full run. I could still hear it following me and talking to me about how I was not supposed to take this... and no one should know about its existence. I was promising that I wasn't taking a piece of it, and no one would be told that it was here. I never turned back around to look behind me as I walked, and I kept expecting it to grab me and yank the bag out of my hand but it quit following me when I stepped out of the tree line and ran into my truck... literally. Yes, I was moving so fast I literally ran right into my truck parked at the end of the trail. I opened the door and hopped in and burned rubber out of there. When I got home I ran straight into the house and locked the door.

I dropped the casting onto the couch and started closing all the curtains so that I couldn't see out the windows, and nothing else could look in. I took the cast out of the bag and looked at it and was very disappointed. The casting material had oozed down into the vegetation and the sand, and we had pretty much lost all distinguishing features. I know now that I should have prepped the sandy soil with hairspray, but I didn't learn this until much later.

My first attempt at casting was a dismal failure, and now I had something that could be up to ten feet tall looking for me because I took a piece of him. I won't lie here; I was really scared being alone that night! I didn't feel comfortable sleeping in my bed because it was surrounded by windows and I thought for sure that this thing was going to come to my house and take his hand print back. It was kind of like the story about the little old lady and the bones. "Give me back my bones...." That sort of thing! I spent that night sleeping on the floor in the hallway. It seems pretty silly now, but at that time it made perfect sense to me. Since I'm still alive to tell the tale he obviously didn't come by to take his hand print back, but that was one long night for me.

Figure 3 Hand Print casting

Chapter 2 – Learning My Way

I stayed in my house for three days after I brought the casting home. I got on the internet and looked up bigfoot. I saw pictures of other hand prints that looked very much like mine, and I read hundreds of sighting reports. I found forums and blog pages and started to read some of the posts on them and I found that no one really had a clue what these beings were or how long they had been here on earth. Everyone had their own theories, but no one could tell me why I was hearing voices in my head and feeling someone else's emotions in my body. In fact, when I shared the things that I had experienced with other sasquatch researchers I found on the internet, they told me that I was either stoned, crazy, or just making things up because this just wasn't possible.

Well, I knew it was possible because it happened to me. It's also a known fact that the military has been using humans for remote viewing for years, which proves that telepathy is possible. Why wouldn't my feeling another's emotions be possible? I decided then that the self-proclaimed experts in the sasquatch field didn't know half of what I'd found out after one encounter, so I quit looking for answers from them. I decided that I had to go back out there and face this thing and find my own answers. I couldn't live in fear of the forest. I had been enjoying it for most of my life and lived to tell the tale. I wasn't going to let this beat me.

I went back out on the fourth day. Since I didn't need them underfoot, I locked the cats in the house and set out. Once again, no birds dotted the landscape. There was no Heckle and Jeckle either, which I found that very odd. I went to the same area and started toward the hillock and I climbed over the other side to the

Bee's nest. I had to see for myself what it really was... It was just what I had thought it was. It was a gall that wrapped around the branch of the tree... not a bee's nest! I felt like I had proven my case on that one, and I was feeling pretty darn good as I walked through the trees and proceeded down the game trail.

I saw that this trail opened up into another large clearing encircled by trees. I spotted what looked to me to be a newly made trail that led towards a very large cedar tree with branches that draped to the ground. I followed the matted grass trail right up to this tree and I noticed that the trail ended right in front of a large branch. I found that odd, but I walked over and I reached out to grab the branch. As soon as my hand hit the branch, it swung open... Just like a door! I was so surprised that I just let it swing back into the closed position and stood there with my mouth open. I couldn't believe it. I had never seen anything like that in all my life. It was a natural doorway into the tree... Not the bole of the tree, but into that area beneath the canopy of branches. Awesome! I swung it open again and entered the tree.

Inside the tree was what I could only describe as a room. The clearance under there had to be six feet high, at least. This was just a guesstimate because I am five feet and three inches tall and I couldn't reach that high. As I stood under this tree looking up, I noticed that the branches had been broken off and my first thought was that this could possibly be a bear den. As I slowly walked around the trunk of the tree, I saw that it was actually two trees growing very close together and forming this little room underneath of them. It was then that I spied the evergreen boughs lying over in one corner.

There were two of them, and they were laying one atop the other, forming what could only be a bed of sorts. The needles were still nice and green, so they couldn't have been there for very long... But that wasn't the surprising part at all. Lying on top of this bed were two bright white, wild bulbs... two perfectly clean white bulbs. There was not a speck of dirt on either one... even the roots and vegetation had been pulled off of them. They were lying side by side on this fresh green bed.

I just stood looking at this for a few minutes. The site in front of me was so surprising and so beautiful that I was momentarily spellbound. I was just standing there wondering what was going on here and how all this had come to be here. Remember, I'm standing under two cedar trees looking at an evergreen bed with bulbs laying on it. In all the time I'd been walking in the woods and mountains, I had never come across anything this strange. I just wasn't able to wrap my head around how a bear could drag these branches in from another tree and lay them down so perfectly, and then go out and pick and wash these bulbs without leaving marks from its claws or teeth, or even any saliva on any of it.

I guess I still wasn't completely sure that the sasquatch could even exist and was still looking to disprove everything I had found up to this point. There had to be a reasonable explanation for all this. And why, in forty two years of life had I never seen, or heard much about these creatures? They couldn't exist. I decided to take one of the bulbs off this bed so that I could look them up on the internet when I got home later. I wasn't sure what kind of bulb they were and I wanted to be able to identify them.

I dropped to one knee in front of the bed. As I reached out my hand to pick one of them up, I heard a voice quietly invade my head and say, "Do not take that. It does not belong to you."

The voice was so calm, and stated this as a fact that wasn't to be argued with, so I just said, "You're right, they don't belong to me!" This made so much sense to me, and it touched me on such a moral level, that I knew without a shadow of a doubt that I would not be taking either one of these bulbs with me that day. I just stood up and decided to leave them lying there.

The next thing I knew a slight wind seemed to push a door open on the off side of the tree. It was just a slight rustling of wind and one of the branch's on the off side of the two trees moved just slightly enough for me to see daylight through that area, making it clear that I could get out that way. I decided it was time for me to leave this place to whomever it was that was living there, and walked through this doorway between the branches. I did, however, think to myself that I would be coming back this way someday to check this out again.

When I exited the tree on this side, I noticed that I was standing with the two mounds a short ways off to my right. I had exited into the second field this time. Was this why I wasn't allowed to come into this area before? So I wouldn't find this bed? And if so, how did I just get in there this time? Did I somehow sneak up on it this time? I just shook my head and spent some time walking around this new field looking for any new evidence and any more tracks that I could find. I never found anything more and I was getting tired so I just decided to call it a day and go home.

The area where these events took place is highly populated in the summertime so I knew that even if there was a sasquatch here in winter, he wasn't living here year round. It had to have just stopped by on its way to somewhere else, or maybe it just stopped here for food. A few days later I began scouring maps for likely places around my home where they may actually live. I spent countless hours, and consumed countless gallons of gas, driving up old logging roads and crisscrossing mountain ranges looking for a more likely home base for these beings. I could have saved myself a lot of time and energy because they were about to let me know exactly where they were.

On May seventeenth, my husband and I were driving down a country road towards the back entrance to a game reserve. We spotted a herd of elk grazing in a very large marshy field, and he pulled over to shoot some video of them. He had to zoom in a long way as the elk herd was grazing way over on the other side of this field. He shot some video of them for a while and while watching the herd we never noticed anything out of the ordinary.

Three days later, on May twentieth, my husband was sitting in the living room watching the video on our big screen television when he called me into the room. He told me to watch the video and pointed to something that he had caught walking out into this field. Whatever it was, it was substantially taller than the elk and was a reddish brown color. It was seen walking into the field from the right side and traveled along in front of the green bushes that bordered this part of the field. It then walked over to a stump and knelt down next to the stump and stayed there. That's when my husband had stopped filming and we got into the truck and left.

He took some still photos from the video and we uploaded them to the computer but when we tried to blow them up in size the picture was too grainy to see any additional details.

We decided we had to go back out to this field and take some new video and see if there was any other explanation for what we were seeing. We didn't see how, but we had to check. Due to the weather and other factors, we couldn't get back out until May twenty eighth. We drove out to the same spot and he started filming the same way as he had before. Nothing on the new video could explain what we had caught the last time, but the vegetation had grown a bit higher by then and we couldn't see the stump as clearly as we had last time. We decided that we would have to hike out into this field ourselves and see exactly what was in the area of the stump.

We knew we could find it again because there is a sign that sat off to the left behind the stump, and it is shining white in the video. We would recognize the stump and the sign, but first we had to find a way into this field since it was private property and we couldn't enter from the road. I don't endorse this behavior, but I was obsessed about getting into that field.

We went back to the maps and noticed an area between this field and where I had my encounters that looked really promising. We would have to park in the cemetery and walk a couple of ridges through the trees and we would end up in the back of this field. While it sounded good on paper, our first trip to this area proved how much tougher it would be. There was a guy camping right where I wanted to walk, but my husband didn't want to walk past him and have to talk to him. I think he was ashamed of what we were doing there that day because he's usually quite talkative.

We walked a different way and saw many bear tracks... Very large and very fresh bear tracks and a lot of scat piles. The only other thing we found that day were a lot of locked gates across every road we found that headed in the direction we wanted to go. When we left, I was a bit discouraged.

I decided to go back out alone two days later. I knew in my heart that there was the way in to that field and I was going to find it! I threw my backpack and my walking stick into the truck and I hit the road. I parked at the cemetery and walked back into this area to find the camper gone. I was heading up the ridge and no one could stop me... Except for what I found blocking the trail... That stopped me in my tracks. I walked up to find a green sapling that had been used to block the trail.

This sapling was bright green and had scratch marks in it. It also looked like it had been hung across the trail. I couldn't see where it could have been growing like this, and it was about waist high across the trail. The scratches ran the length of the sapling and brought to mind bear claws. I stood looking at this for a while, and thought of the gigantic tracks I had just taken pictures of two days earlier and I decided that if mama bear wanted this area she could have it. I decided to check for another way in.

There was another trail farther up to the left that led to another ridge line and I decided to check that one out instead. I walked all the way over there and found there really wasn't a trail in, just some depressions through the vegetation showing where someone had walked through. I followed it for a bit until I noticed that as soon as I climbed over and under this huge fallen tree that stood in front of me, the land was going to fall into a ravine and it was going to get really dark in there! I mean midnight on a moonless night dark in there. So, I scared myself and decided that this wasn't the time of day to check that area out and I would bring my flashlight with me next time.

A funny thing happened then, but I didn't laugh about it until later. I had been immersing myself in all the stories about sasquatch on the internet, and had read quite a few that featured the "Men in Black" coming to people's houses and taking all their research papers and clearing out their computers. I didn't know how to take that one, and frankly, I still don't. But as I was walking out of this area, I had to climb up a large hill to get back to the Jimmy. When I was about three quarters of the way up the hill, I started to hear a very distinct helicopter sound. A deep throaty helicopter too. I kept walking and was looking all over to see where this helicopter could be. I didn't see him in the sky anywhere and it sounded as though it was really close to me.

Just as I reached the top of the hill, the helicopter raised up above the other side of the hill. Just like in the war movies! I quickly scanned the front of the helicopter for guns, but didn't see any. The pilot and I were eye to eye at this point, and I'm sure the look on my face still cracks him up to this day. I was scared, curious, annoyed, anxious, and then scared again, all in the course of a second and every emotion I have shows very clearly on my face. He actually hovered there for a few minutes just looking at me. I was thinking that maybe the stories were true, and the Men in Black were coming for me! I kept wondering how they had found out so fast when I hadn't told anyone what was going on yet? I kept thinking, "Wow, they're good!" He waved, and I waved back, and then he flew off. It turned out he was just the maintenance helicopter for the power lines.

I started begging my husband to go back out there with me. He finally consented and we walked over to the ridge on the left that day since the green sapling was still over the other trail. We followed this trail in and climbed our way over and under the dead tree and then down into the ravine. This trail led us down into this very beautiful and lush area. We stopped to sit on a large fallen log to smoke a cigarette and get our bearings.

While we were sitting there, I was looking around and noticed what could only be described as a woven screen made of branches and twigs. It sat just a short way off the trail ahead of us, and we would have to pass right by it if we continued into this area any further. I mentioned it to my husband and pointed at it and told him that I would like to go over and check it out when we passed by it. He agreed and said it looked like some kind of hunting blind and we continued smoking and talking. We then continued further down the trail and encountered a cross roads.

The trail continued forward ahead of us, but there was also a trail which led uphill on our left, and a trail that led downhill towards the marsh on our right. To one side of this crossing was a broken tree an estimated eight to ten inches around. That is just an estimate because I never actually measured it. It was broken and pushed over about three feet high and the break was relatively fresh. The broken end still had some wet looking wood in it. My husband pointed out the two splintered areas along the bent over portion of the tree. He said they looked more like they had been whacked on repeatedly rather than scratched out because there were no claw marks or teeth marks showing but just the splintered wood. He took pictures of this thing with me standing behind it, and I wandered off looking at all the different sized foot prints around this tree and other places. There were little ones and big ones all around us!

Figure 4 Knocked over tree

We followed the trail uphill first and came to a woven branch wall. That's the only way I can think to describe it... a wall. It was about six feet high and covered the entire trail from side to side. It was set up between two large trees and reminded me of the baby gates we put in between door jambs to keep our kids in or out of an area of our homes. It was created with small tree trunks and then had branches of all sizes woven around, over and through them. It was created very sturdily and it wasn't even easy to look through it... just like a wall.

Until we had walked right up to it, I hadn't even seen it. It blended in very nicely with the surrounding vegetation. We checked it out and decided to continue up the trail anyway. My husband climbed over the top of it, but I followed a set of small tracks over to the left side which led under the anchoring tree and around the obstacle. We followed this trail up the hill for a bit and came upon another wall that blocked off the path. This one was built just like the first one we had found. By now we were getting the message and decided not to climb around this one. We turned and followed another path back down the hill instead.

When we were almost to the bottom of the hill, I noticed that all of the little footprints seemed to be walking right up to a tree and disappearing. Where this tree was growing the land dropped down in elevation slightly, and the actual path veered off to the left of the tree. While my husband followed the path to the left, I followed the small tracks to the tree. When I got there I noticed a branch low enough to grab, so I did. This branch swung me out in the air and around to the front of the tree, where I let go and found more little tracks. It was so much fun! Just like a swing! I turned around and said, "I want to do that again!" but my husband gave me one of his stern looks and nipped that idea in the bud.

We continued down the hill and found ourselves back at the crossroads again, so we followed the downhill trail towards the marsh this time. We found ourselves in a clearing of a sort. It was surrounded by trees and bushes on three sides, with the marsh right in front of it. There we found a very large evergreen bough laying on the ground and spread out like a bed. It still had greenery growing on it. I walked over to it and was impressed by the sheer size and depth of it. It was very deep, with a very clear indentation of something that had been laying in it, and they weren't wrapped up in a ball either. I could see a head indent, a body indent and where a leg had been draped over the side of it.

We took pictures of this bed from various angles until my husband started looking tense and said we should go. I asked why he felt like leaving so soon and told him that I wanted to go walking through the marsh, but he was adamant that we leave, so I decided to save that for another day.

We were about three quarters of the way out when he turned to me and said that something was following us. I asked if he had seen anything, and he said, "Its right over there in those bushes."

Figure 5 Giant Hemlock Bough

He never did say he had seen it, or what it was and, judging by the look on his face, I could see that he didn't want to discuss it either. I then remembered that I still had two bottles of water and two unopened packs of Ritz crackers in my backpack. I had also picked up a blue feather during our hike that day. It was a very large feather and the most beautiful deep blue that I had ever seen. I still don't know what species of bird it came from.

I pulled out the water bottles and opened one up and took a drink and handed it to my husband. I then stated, out loud, "This is clean drinking water. We will share this one, and you can share this one with your family", and put the second one on a large fallen tree that was laying in front of us.

I then took out the cracker packs and opened one. I put a cracker into my mouth and chewed it, and then handed the pack to my husband. I then repeated, "This one is for us to share, and this one is for you and your family to share" and laid the other on the tree.

I then pulled out the feather and said, "I gift this to you, from me, because it is the most beautiful feather that I have ever seen!" I laid that on the tree also.

My husband was looking at me like I just went crazy, but didn't say anything to me. We finished our crackers and the water and put the empties back into my bag and we left.

It wasn't until after we returned home and we were looking through the pictures we had taken that I remembered the blind that was set off the trail. I sat up straighter and asked my husband why we hadn't stopped to check out that blind, and he looked at me with a strange look on his face and said he had forgotten all about it. I told him that I had forgotten all about it too, and we walked past it twice, once on the way in and then again on the way out.

Early the next morning, I went back out to check on the things we had left on the tree. I had been thinking about this all night and wondering if the sasquatch could really have been out there. Did we really see the things we had seen? Was I hoping it was from the sasquatch when, in fact, it really wasn't? I had to get back out there and prove it to myself one way or the other. When I got there I found that the crackers and the water bottle were nowhere to be found. There were no scraps of either one lying around either.

The feather was still there, but it had been altered. The feather had been picked apart from the top down to form three different chevron pieces, and the tip of the feather was gone. I never knew that if you plucked off the top portion of a feather, all the rest came off as chevrons. It was so cool! The bottom part of the feather was still perfect and still intact. It was lying on the tree just as I had left it... not moved up or down at all, and the chevron pieces had all been laid in a pile next to the bottom half of the feather. They had also added a very small, downy feather for me. It was laying with the blue feather. I said out loud, "Thank you! I hope you liked the crackers."

Figure 6 Feather on the log

I felt bad that I didn't have anything with me to leave for them that day. I really hadn't thought they were going to take the crackers and water, and I hadn't brought any snacks with me that day. It was pretty surprising to me that they took the crackers in the first place. I was prepared to think that people were nuts when they said the sasquatch would gift you different things for leaving them food. I had to change my tune about that one and quickly!

No other denizen of the forest leaves gifts for food, and none can pick a feather apart, or place things in a pile without using their mouths. I was going through some major mind changes during this time! I did check the feather for teeth marks and wet spots to see if it was in anything's mouth and the answer was no. Something with fingers pulled that feather apart and set the pieces in a pile right next to the bottom half of that feather. I looked around the area for prints, and for pieces of the bottle or cracker wrapper, but I never found either of them.

That got me obsessed with the entire phenomenon that is the sasquatch. I was now eating, sleeping and living sasquatch, with a passion. I knew I had to get back out there every day to see what was different and what was new. I started pouring over every story I could find on the subject and checked out every book the library had. I joined forums, and read blog posts, and perused all the websites I could find.

The more I read, the more I found that no one knew much of anything about this species. I was still on my own with this. None of the places I checked knew anything about the Twilight Zone part of the sasquatch and what I was experiencing either. If they did, they sure weren't talking about it. I just kept hearing that I was a nut job and one lady even went so far as to say that I was a kook and part of a cult. Not exactly what I would call a warm welcome into the bigfoot community, so I started hiding the strange parts of my encounters and experiences, and decided to just go solo.

Chapter 3 – Further Discoveries

One morning I received a surprise phone call from a man that I had never met. He introduced himself as living in Indiana and told me that he had received my phone number from a lady whom I had spoken to over the internet. He then stated that he had been resting quietly and had been given a vision of me and my research area. This intrigued me so I continued to listen to what he had to say. He asked me to write down everything he was telling me for future validation so I quickly reached for a piece of paper and a pen and asked him to continue.

He told me that the blocked trail I had found was the work of the sasquatch, not a bear, and that the trail would be unblocked for me the next time I went to the area. He said that I would find sign of them being there and to keep an eye out in a wet area for a barbed wire fence. He said there would be hair on the barbed wire fence, and for me to collect that hair. He then told me how many sasquatch were in this area, and what sizes and colors they were. He also said that the clan did not live in this area; they only came here to collect food stuffs during certain times of the year. To be honest, I didn't know what to think about this and took it all with a grain of salt.

I told my husband what this man had to say, and asked him to go back out with me. Of course, he thought I was nuts for listening to this man, but he finally humored me. We went back out to this area and did indeed find that the sapling had been removed! That surprised me because this wasn't something that a person from Indiana could find out over the internet or from a television report. He would have to know someone who lived in this area to know that the trail had been blocked and to know that it wasn't now! I had never mentioned the sapling to anyone in any forum, blog or chat room.

I stood right where the sapling had previously hung and said out loud what this man had told me to say. "I wish to come into your home and I ask permission from you." I then added, "If you don't want me to be here growl, or throw something so I'll know." As we stood here for a few seconds, I risked a glance at my husband... he was obviously thinking I had gone nuts.

As we entered this trail, I noticed immediately how packed down and smooth the ground was. I also noticed how many things we saw that just didn't look like the work of Nature. At one point, we had to duck under a very large fallen tree that blocked the path. I turned around to look at the back side of the tree and noticed two separate places where the branches had been twisted and woven into round, wreath like shapes. It was really cool looking. In a few different spots we found foot prints, and hand prints climbing up the muddy walls along the sides of this trail. That thrilled me to no end!

Figure 7 Twisted trunk of tree

I spotted a medium sized tree growing on the edge of the path at one point that didn't look quite right. The top of the tree had been twisted numerous times and then bent in a downward position, while the rest of the branches stood straight around it. The bend in the branch pointed down into the ravine next to it. What a sight that was. I had never seen anything like this before.

As we started into the ravine below this tree, my husband pointed out a strange imprint in the drying mud. Since neither of us could identify the track, we sprinkled flour around the outside of the print and took a picture of it. I wish I still had that picture for further identification but, unfortunately, this was one of the pictures that I lost in a computer crash.

The early pictures that I did include in this book were the ones that I had posted to a social media site on the internet, and I was able to copy them back onto my new computer. Now I wish I had shared more of the pictures I had taken. That way I would still have them. I did copy them all onto a disk at one time, but with all the moving I did between 2010 and 2012, I can't seem to find the disk anywhere and fear I may have lost it somehow.

After we took the picture of the print in the drying mud, I spotted something that looked like a domed shelter down in the bottom of the ravine. It looked to be made from assorted sized branches and other vegetation, and was set in between two large trees. I was too nervous at that point to walk down the hill and check it out, so I just took some pictures of it from up on top of the hill. I was getting an eerie feeling from being down in this area, and I really didn't want to walk down to the bottom if I could help it. I would have followed my husband if he had wanted to go. Luckily for me, my husband didn't want to go down into the bottom of that ravine, either.

We climbed back out of there and continued walking down the main trail in the direction of the field we were trying to access. We came across a long muddy spot on the trail. My husband had been walking in front of me when he stopped and pointed out something on the ground. As I looked around him, I saw the tracks of elk, deer, coyote… and one perfect barefoot track! It was clearly visible and showed five toes!

It appeared that the sasquatch had been skirting the mud patch and walking through the vegetation on the left of the trail and for some reason it had stepped into the mud with its right foot. It left one perfect print, and then stepped back into the vegetation again. I was so excited to see that print that I got a little more gitty-up-go in my gait. I was seeing more and more evidence that the sasquatch really were in this area, and maybe that man who had called me actually knew what he was talking about.

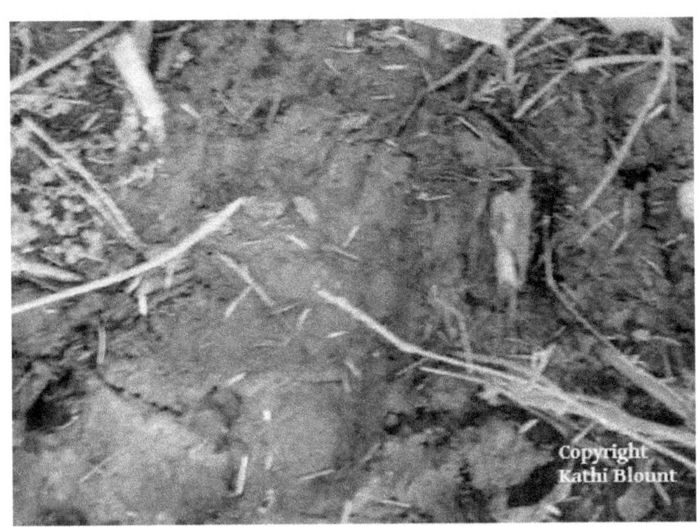

Figure 8 First footprint find

Although we were finding so many unique and amazing things in here, I was starting to have my doubts about finding any barbed wire fencing. I didn't want my husband to be able to say that this man was just a nut job. My husband was the kind of man who would stick you to every detail, and the barbed wire fence was the only thing we hadn't found yet!

After we left the track in the mud and continued down the path a little ways, we came to an area with a small hill on the left side of us and my husband wanted to climb up the hill and look around. Just then I happened to look off to my right and I noticed a trail that was leading down into and then through the marsh. This trail was unique because it was the only high ground in that area, and it cut straight through the marsh. It was obviously being used as a main thoroughfare too because all the vegetation along it was squashed flat, while it was still growing high along the sides.

I walked down to this trail and out over the swamp and I noticed that sitting off to my right side was a branch structure that was built out in the middle of the marsh water. It was made from branches and looked just like a four feet tall tepee. It had a stick balanced across the top that was pointing off in the direction of the trail. Since I would have to step off into some really gross water to get close to it, I just took a few pictures from where I stood. While I was doing this, I heard my husband call to me from up on the small hill.

Figure 9 Field through trees

When I got to the top of the hill and began to tell him what I found, I noticed that he was standing in front of a barbed wire fence! As I got closer, I noticed hair hanging off the fence. I just looked at him and smiled real big. He never said a word. We did take samples of the hair off the fence, but later found out it was from an elk. Well, the guy never did say it was going to be sasquatch hair, so I guess we trumped my husband on a technicality.

We continued to follow the main path and came to a deep stream that was blocking our way. We noticed that a tree had been laid across this stream to make crossing possible without having to get wet. We also noticed that various sized five toe tracks had crossed this muddy tree. I don't know what he was thinking, but I was jumping for joy inside. I was finding proof of this species all around me, and I could no longer doubt their existence.

When we crossed this stream we found ourselves in a lush green meadow and I looked up to see the most amazing sight. The tree line was right in front of us and the sun was streaming through a gap created by two medium sized trees... And through this gap, I could see the field we had been seeking! We had been searching our way into this field for seemingly forever and now we had it!

We had finally found our way in and we had followed the tracks of the sasquatch to get here to prove to ourselves whether or not the sasquatch were ever in this field. It was pretty funny when thought about... We were on a mission to get into that field, and we were accomplishing this mission with the help of the sasquatch.

We really had no idea how large and wide this field was when we first arrived. We also didn't know that there was a river zigzagging back and forth through the entire middle of it with five feet to six feet high muddy banks that were all but impossible to climb. I also started noticing just how many places there were for large animals to hide and we were going to have to walk past every single one of them. I was a bit leery, but that didn't keep us from going in. We had found the field and we were going to get to the other side to check out that stump.

We had to keep following the river back and forth through this field trying to find a way to cross it. We finally came upon a very large tree that was lying across a fork in the river. It was put in the perfect place also... A place which allowed us to only have to cross the river once and we were in the back part of the field exactly where we wanted to be.

When we made it across, we noticed an elk herd grazing among the trees to the left of us. We stood watching them for a few minutes, but I was getting nervous standing so close to such magnificent creatures. I guess I've watched too many "When Animals Attack" shows and I kept waiting for the male to come charging at us. We did eventually see him wander out into the open and nothing bad happened but it was time to be on our way.

We finally spotted the white sign that we were looking for and there was the sun bleached stump sitting right where it should be. We checked all around it from different angles and saw that there was nothing there that would have explained what had shown up in the video/pictures. There was just a sun bleached stump sitting there all by itself. There was nothing reddish brown in front of any of the bushes that this thing had walked front of in the video either. We were convinced that we had filmed a sasquatch! It may just be a blob-squatch to others, but we knew what it is.

We walked back out the way we had come in and I had to stop at the entrance to look for the green sapling that was hanging there when we had seen it before. I just kept thinking that it had to be there somewhere. I was sure the bear didn't take it away, and I didn't know why the sasquatch would be using it. We looked around for a while and just as my husband was saying that we shouldn't bother and should just go home, I decided to dive up under some bushes and look there. I found the sapling pushed up under the bush and pulled it out to take a look at the scratches again. Upon further examination the marks really didn't look like claw marks at all, and I sure do wish that I had taken it home with me.

I felt safe enough out here to return alone a few days later. First I went to stand on the side of the ravine and check out the dome shaped shelter. It was still there and since I still didn't want to walk down into this area I continued on to the muddy spot where we saw the footprint. It was the first foot print that I had seen in the ground and I had brought the casting materials with me this time. I had read about the proper ways to cast prints and I was going to do this one the correct way. However, when I arrived at the spot, there was no print there to cast. The area had been full of deer, elk and coyote tracks also, but these tracks weren't there anymore either! The soil was as smooth as if it were just laid concrete. I just stood there shaking my head, and looking all around to make sure I was in the same area again. Yes, I was in the right area, but where did all the prints go? I never found an answer to this, and I still do not understand it.

I walked all the way to the back, past the stream, and then into the area that opens up onto the field. I had come out here with my recorder to play some music, and try to spend some time out here letting the sasquatch get used to me being around. I was sitting there playing and enjoying the day when I noticed that the birds were singing all around me. I then remembered that the birds weren't singing when we were out here last time. I also remembered reading that the forest will get real quiet if the sasquatch were around so I figured that they must have already moved on.

When I left this time I walked up the small hill where we had found the barbed wire fence earlier, and then down the back side of it. I wanted to find a different way to walk out of this area and check it out a bit more thoroughly. While I was hiking I stumbled upon a very large field of skunk cabbage and I noticed that there were numerous holes in the patch. I walked over to check it out and noticed that the roots had all been removed, but the leaves were still intact and laying on the ground in a circle around the holes. It appeared that the roots were just pulled out of the ground. It was an odd sight and one I had never seen before. I took pictures of this and continued on my way.

I then came upon a track way left in the mud. There were six tracks and each print was barefoot. Each was approximately twelve inches to fifteen inches long. I first spotted the tracks crossing a small stream and then out onto the muddy bank where they continued up onto harder soil. I followed the tracks off in the direction they led for a little while, but then I realized that I didn't know enough about this species to go following one around while I was out here alone.

Before I headed home that day, I walked over to the log where I had left the crackers, the water and the feather previously. I wanted to leave the rest of my lunch for them just in case they came back to the area, and I had brought along a little something that I hoped the kids might like. I left them another package of Ritz crackers, a bottle of water, and a small leopard Beanie Baby toy. I thought maybe a wild cat would be something they were familiar with, and would enjoy playing with. I even took the tags off of it so they wouldn't choke on them.

Chapter 4 – The Cemetery

I had been relaying all my finds and experiences to my girlfriend, Cristy. She came down and stayed for the weekend and I took her out to show her what I had found. First I took her to the area that had been blocked by the sapling and she was amazed at the things we were finding there. This time she pointed out that even the grasses growing along the sides of the trail had been woven together. Some of the designs were simply amazing.

I showed her the wreaths woven on the fallen tree. I think the woven grasses along the trail were the biggest surprise to me. Who would take the time to do all that work? It was an extremely pretty sight to see, that's for sure. I showed her the skunk cabbage patch, and there were more holes in evidence, and more roots missing.

We then went over to the area where I had left the goodies for them. The water and crackers were gone, but the Beanie Baby was still there. We searched around the area and found the water bottle all chewed up with the cap lying off to one side. We continued to look around and found the cracker wrapper all chewed up and scattered around also. I knew then that it was the bears or coyotes who had found these items, and I was pretty disappointed.

The toy was still laying on the tree, so we left it there and started to walk farther up the trail so I could show her the tree with the swing in it, the bed by the marsh and the broken tree at the cross roads. We hadn't gone far when Cristy stopped me and quietly asked if I could see the face staring at us through the bushes up ahead. I looked around but couldn't see anything, so I asked her what it looked like. She said it looked like a brown and black face looking through the bushes. My first thought of that description was a bear, probably sitting there wondering what we were going to do next. What we were going to do next was back out of the area, never turning our backs on the bear, and leave it the heck alone. We didn't breathe easier until we were back in the sunshine and off that trail.

The next weekend my husband and I decided that we were going to go back out to this area. This time we intended to climb to the top of the ridge to the right and see what we could find up there. When we crossed over the fallen tree and started walking down into the ravine, we found that we were following pieces of a torn white T-shirt that had been tied to trees along the different trails. Someone had been trying to mark their way through here. Every time I saw one hanging from a branch, I pulled it down and stuffed it into my backpack. I figured that if they didn't know their way around in here, I sure wasn't going to help them find their way back in. I was already becoming very protective of this clan.

Figure 10 Three equal sized sticks

About halfway to the top, we came across a small trail that appeared to lead straight into a bush. When I got closer to it, I could see a thin spot in the bush where the trail led underneath it. My husband declined to come with me, so I got down on my hands and knees and followed the trail in. It led me into a beautiful little clearing. The ground was covered in moss and various berry bushes grew on all sides of it. It looked like it would be the perfect playpen. Once inside, I looked to my left and noticed three white sticks leaning up against some branches.

Each white stick was the same length and each one was broken, not cut. The bark was missing from these sticks and they appeared to be very sun bleached. They were created by breaking one single branch into 3 perfectly equal pieces. It was quite the sight to see. They were left leaning upon some brown branches that still had the bark on them. This felt so special to me that I had to get a picture of it. I could imagine a youngster standing here and trying to mimic some of the structures that its parents made.

We were finding more of the 'woven walls' all the way up to the top that day but we continued on by them this time. We were on a mission to get to the top of that ridge. We noticed a half-eaten apple sitting on a log at one point, but since we knew that humans had come this way, we disregarded it and continued on our way. We were now walking through an area that had very dense vegetation growing on both sides of the trail. We started hearing something paralleling us through the brush.

We kept our ears open but continued walking at an even pace until we got to an area at the top which opened up into a wide expanse. We made our way to the middle of this area and stood listening to whatever it was that was following us. It wasn't making any noise other than the rustling of the bushes as it moved forward. Moments before it would have come out of the bushes in front of us, my husband reached down and picked up a stick and whacked the tree next to us as hard as he could. It scared the heck out of me because I didn't know it was coming and it scared the heck out of whatever was following us because it instantly turned around and went crashing through the brush in the opposite direction.

I turned to him and joked about how I hoped that it wasn't a young sasquatch running home to tell his daddy on us, but he said it was probably just a bear. He was never in a joking mood when we were out in these woods for some reason, but this got me to wondering about what had just happened. Were bears so afraid of the sasquatch that if one hits a tree, the bear will instantly leave the area? Or was it just that the tree whack sounded like the crack of a rifle? Either way, I was going to remember that trick for future use.

We walked all over that day, but the only other things we found were some animal's bones minus the skull, and what could have been a coyote den under a large fallen tree.

I made an amazing discovery in the cemetery when I returned a few days later. This is a small Pioneer Cemetery located in the country, and the perfect place to park to gain access to the area I was currently hanging out in. I parked here as usual one day and as I was getting my gear out of the truck I happened to look off into the cemetery proper. I noticed that the grave nearest me had a bouquet of flowers scattered on the ground, and I could see the bouquet holder lying on the ground in front of it. Since this sort of behavior really irks me, I walked over to the grave with the intention of cleaning up the mess and putting the flowers back where they belonged.

I picked up the flowers and put them back into the cone shaped holder. I then leaned over the grave marker to place the bouquet holder back into its slot and that was when I saw the white carnation laying there on top of the grave marker. It was a fresh flower, not a plastic one, and it had been bent and shaped into what to me looked like a capital A with an H attached to it. Upon further analysis I saw that it was actually two fresh carnations that were used to make this pattern. I didn't know what to make of this, but I decided to leave it there and I put the bouquet back where it belonged.

When I stood up and looked around, I noticed that this wasn't the only grave that had its flowers thrown on the ground. I walked over to the next grave and bent down to pick these flowers up. This was a bouquet of yellow and blue flowers and they were plastic but laying in the middle of the grave marker was another AH pattern. This one looked the same as the first and was also made from two fresh white carnations!

I took a quick inventory of the area and found that approximately forty percent of the grave markers had the carnations lying on top of them, and each one was made from fresh, white carnations, all in the same AH pattern. I took a picture of the flower pattern for future identification. I wasn't sure if it was a military insignia or what. I never did match it up to any known organizations, but I did find a picture of it on the internet later. A woman from another state had found this pattern left on her car trailer, and it was made from sticks. I later found this pattern left in a campsite I frequented also, but at the time I didn't know what to make of it.

I felt that the sasquatch had to have left the coastal region by now, and I kept pouring over maps of the area trying to pin point where their next stop would be. I also started emailing some of the people I had met on the internet in search of sightings and areas of interest. I was told about a hunting cabin which had semi-recent activity, and was located right across the bay from me. Sasquatch had been spotted in this area and prints were found here also, so I thought it sounded like a good place to start. I drove many miles down logging roads and up spur roads before I finally found this cabin. It was located in a very wild part of the forest and I decided that I would get out and investigate it when I had my husband with me.

We returned to the cabin on a Saturday morning and trekked our way down to the shoreline. It had such a gorgeous view across the bay and I envied whoever owned the cabin. There were a lot of ducks swimming in the surf and eating on the shoreline when we burst out of the tree line and onto the shore, but they immediately flew away in the direction of the game reserve across the bay. Smart ducks if you ask me, they knew where safety was.

We found a very large hand print there on the shore above the tide line. It showed four fingers perfectly. We spent some time checking it out and trying to make sure it hadn't been caused by rocks shifting with the tide or something else possibly being dragged across the sand. Finally, even my husband, the critic, had to conclude that it was a hand print.

We continued walking up the shoreline looking for new evidence. We spotted a picnic table lying on its side in the surf, so we turned it over and dragged it up into the shade to sit down and eat lunch. My husband was sitting there looking farther up the shore line while I was sitting and looking at the steep bank next to us. I noticed that it looked like hands had been ripping out large chunks of the Moss and Lichen that were growing on it and leaving wide finger marks in the mud. I couldn't see any claw marks anywhere and I sat wondering if sasquatch eats moss and lichen or would they would use it for medicine of some kind? I decided to look into uses for these two plants when I got home.

As I was sitting at the table and munching on my all-time favorite food, trail mix, my husband asked if I could see what it was that was sitting between those two trees farther on down the shoreline. He pointed the place out to me and I stared over there for a few minutes but I couldn't see anything clearly. It just looked like a black shape between the two trees to me. We watched it for a few minutes more to see if it would move, but it never did. He took a few pictures of it and then said he was ready to leave. I left some trail mix on the picnic table and we hiked back out.

When we got to the top of the driveway, I saw that a pickup truck had pulled up behind my Jimmy and there were a handful of guys standing around with shotguns in their hands. They asked what we were doing down there, and since I had yet to learn about the sarcasm and insults one receives when one mentions the name sasquatch or bigfoot, I casually told them what we were doing there and what we had found. One of the guys smiled and said that he had seen a sasquatch. I got excited and asked him where he had seen it. He started to tell me a story when all the rest of them started snickering behind him.

I noticed at that time that my husband was standing off to one side and hadn't said a word since we had reached the top of the driveway. Another one of the men then turned to me and said, "He's just pulling your chain, the only sasquatch he's ever seen was a statue." The first man said something like, "What did you tell her that for?" This really irked me, and I don't like being made fun of, so I decided to get even.

I asked them what they were doing there, knowing full well that it was to hunt. One of them replied that this was a private hunting cabin and they had permission to be there hunting ducks. So I turned to him and told him that there had been flocks of ducks down on the shoreline when we got there, but I had scared them all over into the game reserve. "Oh, you can't kill them in the game reserve, now can you? Well, I guess you're done for the day and can go home now! Or… do you want me to run down the street and you can shoot at me?" I'm sure a few of them actually wanted to, but I just turned my back on them and got into my vehicle. I made sure to wave goodbye to them as we drove off. After this my husband didn't seem to want to go out into the woods with me anymore. Go Figure!

My lesson that day was that not everyone wants to know what's out in the woods and would rather make fun of you than to consider the possibility that the sasquatch even exists. After this we had some very severe weather move in and I was kept out of the woods for the rest of the winter due to frigid temperatures and an overabundance of snow.

Chapter 5 – A New Year

Since I had to stay close to home over the winter, I spent the time poring over the many pictures and video's we had taken. I diligently searched the maps for a place the sasquatch might call home. I knew they would need water, plenty of food sources, and a thick forest to hide in. I noticed that in some of the pictures, we had sasquatch watching us. My husband wasn't as convinced as I was about some of the pictures, so the only thing to do was to go back out to each of these areas and take a comparison picture of what the area looked like now. We would check for things that were the same and things that were now missing.

As soon as the weather cleared we went back to the spot where I had taken the pictures of the tepee in the marsh. I wanted to analyze this formation more closely and get better pictures of it. The problem with this was that the formation wasn't there anymore. I couldn't even find any of the sticks that had formed it! I found this to be very strange. Most branches and things that are lying around in the forest don't just disappear, or walk away, on their own but the only things in this marshy area now were skunk cabbage plants.

After looking around this area for a while and finding nothing, we continued on and found a tree that had the bark peeled off one side of it. It was a very tall tree and the bark was peeled away from higher than either one of us could reach. There were no claw marks visible on the inner bark at all, it was nice and smooth. I don't know what caused this, or who did it, but I mention it because it was one of the things we found that day.

We came upon another skunk cabbage patch with holes where the roots should have been, and a medium sized skeleton without a skull, lying in the middle of it. I wanted to find the skull so we could identify the animal, but we never did.

Figure 11 Bones and skunk cabbage patch

At one point we had to cross over a fallen tree to avoid a large pool of deep marsh water. As I was crossing the tree behind my husband, he suddenly asked me, "Did you see that?"

I had been looking down and watching where I was placing my feet so I hadn't seen anything at all. When he asked me that question, I looked up at him really fast and lost my footing on the wet bark and fell onto the log. My reading glasses flew off the top of my head and I heard a deep growling sound … HaHeHa … and it sounded like someone laughing.

That irked me and I either asked him what he was laughing at, or said don't laugh at me, because he snapped back at me and told me he wasn't laughing! I didn't know what to say to that, so I just asked him what it was that he saw.

He said, "It was a large black shape and it just ran behind the root wad of that big tree over there."

I asked if he thought it was a bear and he said it ran on two feet. I just slid off the tree and tried to find my glasses. I began to get irritated because they had just fallen off my head so I knew they had to be laying there somewhere, but I couldn't find them anywhere. I finally just gave up and we continued on our way.

I wanted to head for the bed we found in the clearing and check on a few things. For some reason my husband kept trying to get me to go in the wrong direction. We argued about the correct directions all the way into this area and I finally just walked away from him and to where I knew the clearing to be. When we got there we found that the bed was no longer in the clearing either. After walking around a bit we also had to admit that the branch wasn't even in the immediate area anymore. He kept trying to say that this must be the wrong area, but I knew it wasn't.

I pointed out the large tree that the bed had lain under, the large tree on the outskirts of the clearing, and the fallen tree in the back of the area. All looked the same as in the photo, so we were in the correct area. What he wanted to think was just a large growth on the side of a tree wasn't growing there now. Something had been looking around that tree and watching us! What looked like a stump under the dead fall tree wasn't there now either. So it couldn't have been a stump, unless someone had come in over the winter and removed it from this spot without disturbing any of the other vegetation or dead trees. Not one of the questionable things that showed up in the pictures was here now. We had been gifted with pictures of the sasquatch.

Remembering how I described this clearing as having a marsh on one side of it... Now the marsh was pretty much dry ground and we walked out in that direction to see what may be out that way. After walking along for a while and not really seeing anything interesting, we sat down to smoke a cigarette. My husband sat facing me and the short tree line that was located a ways behind me while I was facing him and the marsh behind him. We were just sitting there enjoying the early spring sunshine when he asked me in a very small voice if this place had ever burned in the past. He never even turned his face towards me while asking me this. He just kept staring at a spot behind me.

I replied that I didn't know for sure, but I didn't think so or we would have seen burned trees. All of a sudden the blood just drained from his face! It started at his forehead and washed down the front of his face leaving him looking as white as a zombie or a ghost. I was startled to say the least.

He started stammering out incomplete sentences to me. I will paraphrase here to save from offending by his most colorful language. I asked him what he was looking at and he said, "I think it's a burned out stump. "

NO!" he said with a louder voice. "The stump just moved to its left… It moved to the left! Wait, stumps don't move! How can it move, it didn't move!" He kept stammering about how stumps don't move and I started worrying about his sanity. He was looking a bit panicky around the eyes and I'd read stories about how very macho men just lose it when they see something big enough to rip their heads off out in the woods. I wasn't sure how this was going to end since my husband was a logger and considered himself to be very macho. I also didn't want this thing to bolt off in the other direction just yet, so I told him to please stay calm and not scare it away.

He was just sitting there as white as could be and his voice was getting higher in pitch so I figured I'd better see what he was seeing before it left. I stood up slowly and turned to look in the direction that he was looking. All I saw was a fleeting glimpse of something large and man-like taking three steps backwards into the tree line and disappearing as the branches folded back into place around it. It all happened so fast I honestly couldn't tell you exactly what it looked like. I decided that I was going to go on into that tree line and see what I could see.

I told my husband that I was walking over there and checking it out and he said he wasn't coming. I told him that I didn't care, I was going, and I set off across the marsh. I stepped off into water at this point but I didn't care how wet I got... I needed to see what was over there. About halfway across I stumbled on something and looked down to find a barbed wire fence. Was this the wet place where I was supposed to find the barbed wire? Perhaps we weren't in the correct place last time. I let my mind process this as I continued walking to where this thing had been standing. My husband had joined me by this time. I don't know if it was to protect me, or to protect him, but I did feel better having him there.

We discovered that the trees were growing on a small island in the middle of the marsh. The island wasn't big, only large enough for the small grove of trees that was thriving there. The ground was littered with pieces of fir cones. The ground everywhere was covered with a thick carpet of them. No whole cones were in evidence, only these tips. Perhaps the sasquatch was peeling the cones to get to the seeds inside? My husband said it was the work of squirrels and chipmunks. That made me smile because here was a perfectly secluded spot that the sasquatch could hang out in, and I'd read that they like to eat squirrels and other small rodents so they had various food supplies in here.

There was a trail leading back out into the marsh on the backside of this island and I figured that this was where the sasquatch had to have made its escape. I wanted to continue following this trail, but my husband flat insisted that we were not going to follow it back there. He was very adamant about this and judging by the way he was looking at the time, I didn't want to press the issue. He said it was time for us to get out of the area and turned to walk away. I decided that it would be for the best at that point if I just followed him.

He set a very fast pace for us hiking out of the area that day and I was getting so tired following him over the dead falls and climbing the hills that I actually fell at one point. I found myself on the ground in the mud and on my back with my walking sticks lying under me. I had them attached to my wrists at the time, so when I fell on top of them I couldn't move my arms and I found that I was pinned down on the ground and unable to get up. He thought it was funny, but I got mad at him for just standing there staring at me and not offering to help me up. That ended the trip on a less than happy note, and I decided that I would forgo the walking sticks in the future.

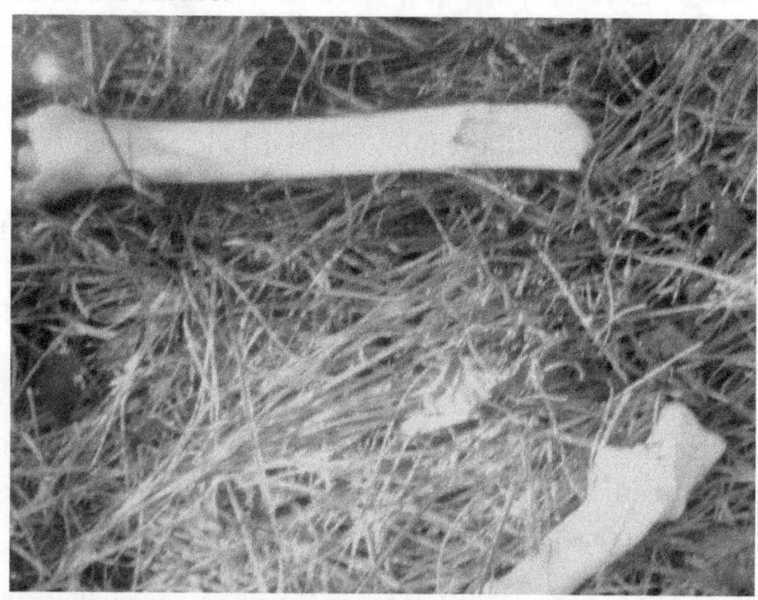

Figure 12 Broken Bones

A few days later my husband came home from work and told me about some otters he had seen playing at the spillway. He had pulled over to watch them playing in the water, and then decided to take me out there so I could watch them too. This dam like spillway was created to keep a large creek from topping the roadway. After going under the road the water spills out into a small creek that empties into the ocean. We parked and started walking down onto the shoreline and came upon a very long trackway that clearly showed five toes in each print. It came out of the tree line, walked through the sharp beach grass down into the water, and then came back up to the trees again. This ground would cut up the feet of normal humans if we tried to walk that distance through it. We took pictures and measured the prints. While walking back to the truck we found a pile of bones at the tree line. Once again, the skull was missing so we couldn't readily identify them. It did look as though the bones were broken and not chewed on.

We returned to this area the next day because we wanted to walk farther down the shoreline and into the trees. We found ourselves in a very beautiful spot where the trees thinned out and lush green grass grew. At first it seemed so serene and beautiful as we walked through here. We came upon a pile of pheasant feathers lying on the grass that appeared to have been plucked from the bird. I gathered up most of these to use for crafts.

I don't remember exactly what we were discussing when we left this pile of feathers, but we started noticing just how many excellent hiding places there were out there for something big to disappear behind. With all the large dead falls and huge root wads everywhere it started looking kind of creepy to me. We then came upon a second pile of bird feathers. These were smaller feathers and included the downy feathers from the bird. Both of these piles were just that, piles. Not scattered around as if they were spit out, but piled one atop the other.

We decided to leave this area since it was getting late, and as we were making our way past one of these excellent hiding places we both heard a low growl. I looked at him and he looked at me and we both started walking faster, keeping an eye open behind us just in case something decided to make a run for us. I still don't know what it was that growled out there, and I never went back to this area again.

One day the wife of one of my husband's coworkers stopped by my house and began to speak of the sasquatch and the things that we had found in the area behind the cemetery. I knew then that my husband must be talking to people at work about what was going on out there. She expressed an interest in hiking out here with me, so we jumped in the truck and headed out. I don't think Lila actually realized how far out we were going to be hiking, because we only made it as far as the entrance to the ravine when she said she was tired and wanted to walk back out. I kept telling her that we only had a short ways to go now. We were standing near the edge of the ravine while we were having this conversation and she was facing me and I was facing the entrance.

While I was trying to talk her into continuing into the area so that I could check on some things, I noticed a bright flash of light coming from inside the tree line behind her. I instantly said, "Someone just took my picture!" This flash of light instantly reminded me of the old time flashbulbs we used to have to place on top of our cameras. It was that bright, and that quick. I saw just one flash of light, and then nothing. This was during a bright sunny day, so the light had to be pretty bright for me to see it.

I don't like the idea of someone taking my picture without my knowledge, and I mentioned this to her as I started to walk over in that direction to find out what it was that I had just seen. She grabbed my arm and asked where I was going, and I told her that I wanted to go and see what that light was and see who just took my picture! She told me that she thought that was stupid, and why would anyone be out here taking my picture. I asked her what she thought the light was then, and she replied that it was probably a reflection off of something. I then asked her why it only reflected once, and why wasn't it reflecting now? I then told her that the ground here goes straight downhill from that tree on, so whatever it was had to be pretty high up off the ground, and I'd been in that area a time or two and didn't think it was a reflection off of anything.

Lila became downright rude to me at that point. She filled me in on the fact that no one believed that the bigfoot even existed, or that I had really even seen one. She said that even my own husband thought I was crazy, and they worried about my sanity. My only thought at that point was getting her out of this area, and then getting her as far away from me as possible, so I never was able to walk in there that day and see what that flash of light was.

However, I did see this same flash of light a few years later while I was attending the Beachfoot gathering on the Oregon Coast. I had walked up a mountain with three other ladies on a night walk. When we arrived at an area of the road that flattened out, we stopped and stood on the road talking to each other. This stretch of road had hundreds of wild flowers growing along the side of it and I just happened to be looking towards them when I suddenly saw what I can only describe as another flashbulb going off. The light appeared and then disappeared just that quickly, and was as bright as an old fashioned flashbulb!

I quickly asked everyone else if they had seen that light just then, but since they were all standing with their backs to it they hadn't seen anything. I told them that I had seen it previously on a sunny day and it was just as bright. One of the ladies said that they were fairy lights, and that I had fairies around me. I wasn't really open minded enough to start thinking that way, but I filed it away for future use. Like I said, I no longer discount anything off hand anymore just because I haven't experienced it yet.

We continued up the mountain a bit more and sat and rested. I took my hat off and left it laying there when we hiked back down the mountain and into camp. My friend Thom was kind enough to give me a ride back up the mountain the next day to retrieve my hat, and I asked him to stop at the spot where I had seen the flash of light. I walked to the side of the road with the intention of walking out there to see if I could spot any metal, or find any other reason for the flash, and found that there was no ground there. We were standing at the top of a very deep ravine!

Once again I had witnessed a very bright flash of light above a ravine, which tells me that whatever it was that flashed, it had to be floating pretty high in the air at the time! It wasn't a reflection off a sign, or a piece of glass, especially the one I saw at night, because there was no light source for it to reflect back at me. I have no idea what these lights were, or why I was the only one to see it each time. Just one more strange occurrence for me to ponder on my journey.

Chapter 6 – New Friends

I became good friends with the man who had contacted me by phone. We spoke often and he always asked me to write down everything he was telling me. That way I could look at it later, and he wouldn't be misquoted. I began to call him Oldie One, so that is what I will call him here. He took me under his wing and started mentoring me in the proper way to interact with the sasquatch. He taught me to respect the sasquatch in all ways, and to also remember to respect all peoples. He told me to think of the sasquatch as another clan of humans and that they have a language, and a culture. He said they lived in family units and raised their young in much the same way as we do.

He always said that he couldn't give me details as to how he knew these things, or give me detailed answers to my questions, because of a nondisclosure agreement he had signed. But I understood that these were truths, and I was to research them for myself if I wanted to know the details. He challenged me, and he guided me, and for that I will be forever grateful. He shared his knowledge and stories with me. I took the time to listen to this great man, and it really helped me to know what to do, and what not to do, when I eventually met up with the sasquatch.

I spent most of my evenings listening to every internet radio show that I could find on this subject, and I even helped co-host a radio show with Oldie One for a few months until I found that I wasn't thick skinned enough to be that noticeable to the Bigfoot Research Community. Being thick skinned means being able to take mean minded criticism and to then deal respectfully with all those who are shouting for your head on a platter because you actually had the nerve to say something they didn't already know, or didn't already agree with. I have a tendency to shout back at these people and that really gets you nowhere at all. It just gives them more ammunition to throw at you and I figured this must be the reason why none of them really knew the truth about this species. How are they going to get to the truth when jealousy or competition keeps rearing its ugly head? So I decided to ignore all of them and just stay with the few people who had not steered me wrong.

Oldie One introduced me to Dallas, a special man at this time who showed me his photos and taught me that I could take photos of them by keeping my camera at waist level and never putting the camera up to my eyes. He stated that when they see you lift something to your eyes, it may look like you're pointing a gun at them and are intending to shoot them. Hey, it made sense to me! He was full of common sense statements and showed me that the easiest ways are always the best ways. I quit carrying around all the high tech gear, and started having wonderful interactions with the sasquatch thanks to this man, and I no longer wanted to try to prove they existed to people... I just wanted to have my own memories of them being close to me. He also taught me a lot about native ceremonies, and shared some words to say to them when they were around me. This man touched my heart and I really enjoyed our conversations and emails.

During this time, I also met a very sweet man who I nicknamed Home Boy, because we were both born in the same town. He was getting clear video and pictures of sasquatch from in the Alum Rock hills. These were the same hills that I used to run into to be alone as a child, and camped in well into my twenties. I never even knew the sasquatch existed at that time. This blew my mind and I still don't know why they never showed themselves to me then. I was able to walk those hills alone during my preteen and teen years on a daily basis and I was never bothered by bear or cougar like everyone always said I would be, so who's to say that they weren't actually there protecting me then? Stranger things have happened to me. This man took one of my all-time favorite videos. Not many people can get a clear shot of a baby sasquatch and live to tell the tale. You rock Home Boy!

And I was very fortunate to be introduced to a man who doesn't deserve any of the flack he takes because of the experiences he has had. I've read remarks about this man that just made me sick to my stomach and all because the people who have had the least experiences are usually yelling the loudest. This man deserves a lot more respect than anyone gives him. I owe him more than I can ever repay and it isn't because of material goods. It's because of emotional and spiritual goods that he has delivered to me. I thank you my friend!

I mention these people here, and thank them here, because here is where they fit. This is the part of my story where I was placed on a different path. I was meeting people who were nothing like the people I was currently surrounded by in my life. These were good people. These were loving people, who were offering that love to someone they had never met in real life. This was a whole new way of thinking for me, and I actually began to feel love in my heart for people I had yet to meet face to face. What a concept that was for my brain to handle.

I began to change. It was a lot like the story of the Grinch. I think my heart grew two sizes at this point. I'm not joking about this either. It happened without my knowledge, and it happened without me noticing. I suddenly had no desire to be around the hard-hearted people with whom I had been associating. I wanted to be around people who saw the beauty in the little things... people who were capable of loving without wanting something in return. I wanted to leave the angst and pain behind, and truly enjoy life and all it had to offer me. I had truly begun to change for the better.

I was talking to one of my new friends on the phone one day and he told me that he was sending me a map that would show me a certain area he wanted me to go to. I finally got my husband to agree to the outing and we set off on a beautiful sunny day. It was about an hour away and the scenery was gorgeous. When we arrived at this site, we had to drive over a quaint little wooden bridge which led onto a one lane dirt logging road.

Figure 13 Two bent trees

My husband pointed out that no other tire tracks were visible on this road, so we must be the first persons to drive up here since the weather cleared. We noticed a spur road that was blocked by a tree which hung over the road and we parked there so we could get out and stretch our legs and walk around a little bit. That was when we noticed that this blockage was actually made from 2 trees that were growing close together. Both of them had been bent over the trail, but the tree trunks weren't broken, just bent down over the road. I had read reports that described the same sort of bent, but not broken, trees and I hadn't quite understood what they were actually describing. It made more sense when I saw it with my own two eyes. Both of these trees still had green vegetation growing on them, and they hung about two feet off the ground. The tops of the trees weren't being held down by anything.

We were discussing the fact that these trees grew on the edge of the road, so snow could have bent them like that. A short way up this spur road we found another tree that had been bent down at ground level but this one was actually laying on the trail. The fourth tree we found made us think twice about the snow bending these trees. It was a very large one that had also been bent, but not broken, and it was very fresh. This one was actually sitting in the middle of other trees up under the canopy, and it was large enough that it spanned the entire spur road when bent over. It hung about four feet above the ground and still had that great pine scent aroma coming off of it. My husband found three separate five toed foot prints at the base of this tree in the dirt. While my husband was checking out the prints, I walked up to the middle of the tree where it hung over the road.

I grabbed the trunk with one hand and found it to be nice and springy like the Johnny Jump Ups that we hang in the doorway for our children to bounce in. I wrapped my arms around it and hung off it. It began bouncing up and down while I hung there. It was so much fun, and I thought about how this would make a great forest swing set! This spur road ended at a very deep ravine, so we walked back to the truck and continued driving up the main logging road.

As we were wending our way around a large hill my husband suddenly stopped the truck. He turned to me and asked if I had heard that sound. I reached over to turn down the stereo and as I did I began to hear a funny engine sound. It was a metallic grinding type of sound, so I instantly thought it was our vehicle breaking down. I asked him what was wrong with my truck, and he told me there was nothing wrong with the truck and leaned forward and turned off the engine. The noise continued, so I knew then that it wasn't the truck. We sat there trying to identify the sound but couldn't. Since we both had recorders with us we decided to get out of the truck and try to find out where the sound was coming from and get it recorded.

It was a throaty, growling, deep metallic sound that we were hearing, and it got louder as soon as I stepped out of the truck and onto the road. This sound seemed to be coming from the adjacent hillside. I stayed near the truck while he started walking down the road. I yelled to him that the sound was coming from right in front of me, and he yelled back that the sound was coming from right in front of him. While he continued walking down the road, I stood quietly for a while and just held out the recorder towards the hillside so we could listen to the sound later when we got home. At one point I heard a rustling, and a sound I couldn't identify, coming from on top of the hill above me and I quickly glanced up, but I couldn't see anything. I was standing too close to the hillside to be able to see over the top and the road wasn't wide enough to allow me to back up and get a better vantage point.

I called down to my husband to see if he had heard the rustling sound and perhaps had seen what was up there, but he said he hadn't heard it at all. Maybe it was because whatever it was that was up there was right above me, and he was down the road towards the other side of the hill. I stood quietly and listened for it to move again, but I never did hear anything more. The metallic growl had never ceased or changed the entire time we were standing out there. I started to walk down to where he was only to find that he was coming back towards me. We just couldn't find one spot that the sound was emanating from. It seemed to be coming from wherever we were at the time. It seemed that the entire hillside was making the noise.

He decided that it was time to leave and wasn't taking no for an answer. We had to continue driving around the hill to find a spot to turn around. I got a good look at three different sides of that hill that day and I saw nothing that would explain the noise we were hearing. And yes, we could still hear the sound all the way around that hillside! There was no logging machinery evident anywhere, or any other machinery that we could see for that matter. I have never heard that sound before, or since, this episode.

We drove out on the same road that we had driven in on, and ours were still the only tire tracks showing in the dirt. Right before we got to the little wooden bridge, we spotted some yellow daffodils lying in the middle of the road. We stopped and I got out and found six perfect daffodils lying directly on top of my tire tracks. The stems were broken, not cut, and the flowers themselves were still perfect. No broken or crushed petals at all! The most intriguing thing about it was the way they were laying there. It looked just like when you are holding a bouquet of flowers in your hand and then you just let them fall. They all fall together and still have that bouquet shape to them. That's what these looked like. Just as if someone had opened their hand and dropped a bouquet of daffodils on this dirt road!

No other cars had driven in on this road, and there were no shoe tracks left on the road anywhere. I then remembered that we hadn't seen any of the usual signs of humans even being in this area. No garbage was lying on the sides of the road, and no cigarette butts were lying anywhere. This led me to believe that this spot wasn't a known hang out for people, so how did the daffodils get there in the short time that we had spent driving over this mountaintop? Could a young sasquatch have picked them for his/her mother? It was a week or so before Mother's Day at this point and I had no idea if they celebrated such things or not, it was just the first thought that crossed my mind. I didn't really know what to think about this find, so I just picked them up and moved them off to one side of the road.

While I was doing this, I spoke out loud and said that I was going to move the flowers to the side of the road so we wouldn't run them over on our way out. Tell your mother hello for me. I felt silly saying this but since these flowers had not been there when we drove in, I knew they had to have been placed here within the last hour. Chances are that we had driven around the corner and scared the poor youngster and he/she had dropped this bouquet and ran for the woods. I felt like maybe they were still there watching us to see what we would do with them.

When I got home I received a phone call from the man who had given me the map. He actually asked me if I had picked up the flowers! I was stunned and asked him how he knew about the flowers. He said that they had been left for me, which is why they were laying in my tire tracks. That way I would be sure to see them when I left the area.

I asked him if he was sure about that, and he replied that of course he was sure. He asked me if I could think of a better place for them to leave the flowers where I would be sure to see them. I thought about this for a few seconds and I felt bad about not taking them. I told him that I had thought they were for a youngster's mommy, so I had left them by the side of the road. He told me that I should be sure to pick it up the next time I found something out of place like that. I felt so bad about not taking them, and I hoped the sasquatch understood why I didn't.

When we listened to the recordings we had made, we couldn't hear the metallic sounds on them at all! We know that we were recording and the recorders were working correctly because at one point we heard a plane fly overhead and that's on the recorder. We could also hear insects buzzing around and our own voices trying to locate the sound. So, we know we were talking about the sound, and we could hear it in real time, but we had no proof on the recorders that it had ever happened. I'm still not sure what to think about that one.

Chapter 7 – Summer of the Clan

I had been given a few landmarks to look for when searching for the sasquatch's likely base camp. I needed to find a spot that offered an abundance of game and edible vegetation. I would need a clean water source for drinking and fishing. Not only that, but it was believed that the sasquatch traveled by way of the rivers and streams. There should be few, if any, people who inhabit the region, but they are seen a lot around church camps, youth camps and horse camps. I was told to look for power lines, quarries and cemeteries. I knew what to look for, and set out to find the perfect place. I found a lot of really nice spots to hang out in, but I never saw any sign of the sasquatch.

One day while I was pouring over maps of the region, I spotted a place that fit all the requirements. It had everything near it that I was looking for and I set off to find a way into this area. However, this turned out to be easier said than done. I kept running into locked gate after locked gate up in these mountains. I kept at it and never gave up trying to find that one single gate that wasn't going to lock me out, but I never did find the yellow brick road that I was searching for.

Summer was coming quickly and I still hadn't gotten any closer to the area I wanted to be in and I was beginning to stress. I had planned on spending my summer getting to know this species, not driving over every mountain top around looking for them. I never did want to ask directions of anyone either, not after what had happened the last time I opened up to someone who lived around here. Once again, fate stepped in and led me to where I wanted to be.

In late May, a couple of new friends mentioned that they were going camping and asked if we would like to tag along for the weekend. My husband wasn't as excited about the idea as I was, but I jumped at the chance to see a new area and go tent camping again. Since we'd been married the only thing we'd camped in was the motor home, and that is not the same as sleeping in a tent under the stars at all! I packed all the gear we would need into the Jimmy and we followed them down the road.

I was completely shocked to see that we were heading right for the exact area I had been trying to access! Things just kept falling into place for me, and I loved it! We had to pass through private property to get there, which is why I hadn't found the road prior to this. I never would have thought to turn in through here.

The higher up this mountain we drove, the more excited I got. My husband warned me not to say anything to them because the other couple we were going to be camping with either didn't know about the sasquatch or didn't believe

Figure 14 Forest

in them. I can't remember now exactly what the reason was, but I knew that I had to keep my secret. We didn't reach the campsite until after dark, so we all ran around setting up camp and retired soon afterwards.

I awoke the next day and fell out of my tent to find myself in a very stately forest next to a very beautiful little lake. The sun was shining brightly and I was so excited to finally be on this mountain top that all I wanted was to just get away from everyone and spend some time hiking alone. Everyone wanted to go down to the lake and do some fishing anyway, so I grabbed my gear and wandered off. I simply cannot describe how peaceful and serene this area really is. The trees grow tall and proud and the branches are so high you can barely see them. There isn't any undergrowth to speak of so it is very open and easy to hike through.

The further I hiked, the more peaceful I felt so I really didn't notice how far from camp I had traveled until I started getting tired. I came across a small babbling brook which was flowing down the hill and I sat and listened to the water talk to me for a bit while I rested. Everything was so cool and mossy under the trees that I was very reluctant to leave that spot at all but it was getting to be afternoon by now and I still wanted to do some exploring before I had to return to camp. I followed the brook down the hill and was just about to round a bend when I felt instant fear.

I was simply terrified and could see no reason for the sudden emotional change. I couldn't figure out why my mind would want me to think I was afraid when just a fraction of a second ago I was feeling so much joy and serenity. Nothing had changed. I have never felt this way in the wilderness before, and this feeling was way beyond any fear I had ever felt prior in my life. I've looked down the barrel of a gun pointed right at me, and even then I didn't feel fear like I did on the trail that day.

I couldn't understand what was going on, but it did make me think about how I had just left camp that morning and hadn't told anyone that I was leaving, or which direction I was headed. That is never a good thing when you're hiking alone, so I decided I'd better head back to camp anyway. I really wanted to know where this path would lead me and decided I would come back this way when I had a hiking partner with me. I turned back and headed towards camp and I noticed that the further from this area I walked the more the fear left me, but I still had no idea what had caused it in the first place.

The next morning my husband woke me up at about five am and told me to get up quietly and not to wake anyone else because he wanted to show me something he had found down the road. I pulled on my hiking boots and followed him out of camp. We walked quite a ways up an old logging road and when we were out of earshot of the camp I asked him what he had found. He just smiled at me and told me to just wait and see.

The road we were walking had deep ditches cut on both sides of it and at one point he stopped and pointed into one of them. I walked over to take a look and in

Figure 15 Print in the mud

the mud was a perfect barefoot print. It sat right below a drainage tube that ran under the road, which kept this spot perpetually muddy. It appeared that whatever had made the print had stepped off the road and left this print in the mud, and its next step had put it up on the forest floor under the trees. The ground there was covered in hard packed soil with needles and vegetation covering it. We could see an imprint of the foot, but it wasn't clear enough to bother casting it.

The print in the mud was very fresh and I think perhaps this sasquatch saw my husband walking up the road and stepped off to hide in the trees. We took measurements and pictures of the print and talked about the possibility of casting this one. I wanted to do it right this time and not have it turn out as bad as the hand print had.

I was so excited to see this perfect print. It even showed the mid-tarsal break and all five toes. We had to go back to camp to get the casting material and water to mix it with before we could get down to business. I didn't want anything to happen to this one, or have the sasquatch come out and destroy it while we were gone, so I wanted to stay behind while he went for the truck. He reminded me that if he walked into camp alone everyone would want to know where I was and what we were doing. He was right about that one; we both had to walk back to camp.

I stood on the road above the print and said out loud, "Could you please not erase this print, or mess it up? I would really like to hang this on my wall so I can always remember you." I didn't know if any of them were even out there still, but it never hurts to ask.

We hurried back to camp and tried to act nonchalant about why we were jumping into the truck and leaving so quickly. We drove back down the road to the print and I instantly jumped out to see if it was still there. It was there... and still in perfect condition. We prepared the casting material and poured it into the print very carefully. We then had to stand around and wait for the material to set up. No way was I going to leave this print in the ground for someone else to pick up. I planned to sit there all day if I had to. While we were waiting for the material to harden, we went ahead and took a look around the area for any additional prints. We never did see any, but we found the sweetest little pond with a beaver dam in the middle of it. I planned to come back to this spot at some point when I could just sit and enjoy it.

When the casting had hardened enough to lift it out of the ground, we wrapped it in a shirt so no one at camp would know what we had. It had turned out excellent. We did a much better job casting this one and it showed every detail of the print.

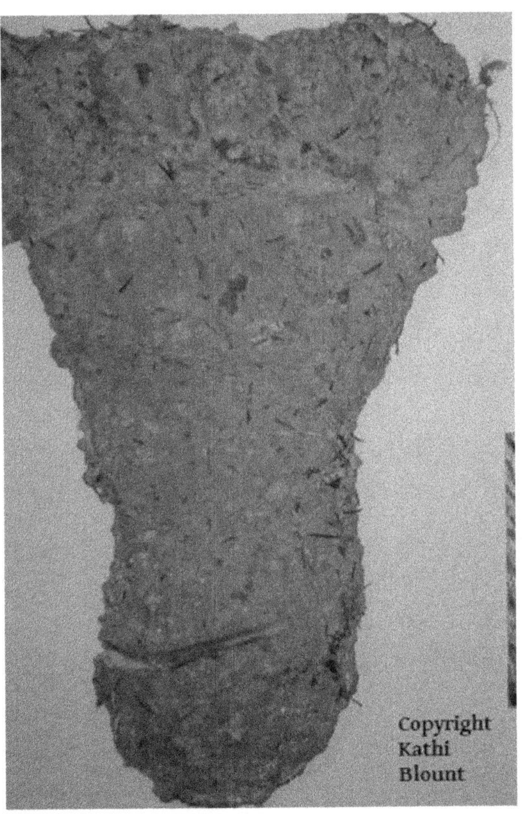

Figure 16 Footprint cast

Of course when we arrived back in camp everyone wanted to know what we had been doing, and of course my husband gave in and told them. Then everyone wanted to see the casting, and he showed it to them. I then spent the next half hour or so listening to snide remarks and trying to answer questions I couldn't answer.

One of the remarks that I heard was, "They killed that one in the Sierra Nevada's didn't they?" When I heard this I just nodded my head and said, "Yes, they did kill that one, and no, sasquatch doesn't exist." It just seemed easier at the time. I wasn't ready to take on anyone with facts that I just didn't have yet.

At one point on this trip, my son asked me to hike down to the other lake with him. I really didn't want to because we would have to cross the clear cut area to get there. I had been told that there were a few very large reddish colored males living up there that were none too happy about the clear cutting and didn't like humans at all. I had been given explicit orders not to go into that area, in fact. I didn't know if this information was valid or not, but I really didn't relish the thought of wandering through that area. He finally talked me into going with him by telling me that my husband had shown him the quickest way to get there and we would be down on the other lake in no time.

This, however, was not the case. The entire area was littered with dead fall, and huge downed trees. He told me to just climb up on one of the trees and we could walk the trees all the way to the lake. Well, this may work for someone with long legs, but it sure wasn't working for me. I had to keep climbing up onto fallen trees that still stood higher than I am tall. Then after walking along the length of this tree, I would have to climb back down and then climb back up onto the next downed tree and so on and so on. After about a half hour of this, my back was killing me. Not only this, but I began to notice that he wasn't headed in a straight line at all. We were zigzagging our way around this field in circles, and I kept thinking that at any moment one of these irate males was going to come charging at us to get us out of the area.

It was quite stressful, and my back was screaming at me. At one point I just broke down into tears and told him that I couldn't do it anymore. Of course he had to chide me and start making fun of me and showing me how you just jump from one log to the other! He made it look so easy, but I just knew that my body wasn't going to cooperate like that. At that point, he jumped off the log he was standing on, which was still a ways away from me, and looked up at me and said, "Mom, come here. Look at this track!"

I asked what kind of track, and he said a large footprint! Well, that was the turning point for me and I told him, "That's it. I'm going back to camp now."

He was kind of mad at me for not following him to the lake that day, and I felt bad about it, but my back was hurting and I didn't want to be in that area in the first place so I walked back to camp to relax my back.

I wondered if this was the same group that hung out by the cemetery. I didn't know for sure, but I planned to spend a lot of time up here trying to find out. I also had to bring my friend Cristy up here with me to check this out for herself. I had tried to describe the things that had happened to me and she had seen the woven grasses and the woven branches, but I knew in my heart that she still didn't believe any of it. She may have wanted to think that they were real, but she needed some proof of her own and I figured that this was the place where she was going to get it.

Chapter 8 – Back Up the Mountain

We planned another camping trip for the next weekend and this time I invited Cristy to come along with us. We convoyed up to the lake in four different vehicles and this time we pulled into a different site to camp in. This camp spot was a wide open space, surrounded on all sides by the thick forest. I figured this was going to get pretty interesting after dark. Since the site was so large, we were able to scatter our tents farther away from each other, and my son was staying in his truck out by the lake.

The spot we had camped the first time was a small circle that was completely enclosed by berry bushes with a small trail leading into the spot. It was very isolated and the sasquatch would have to stay behind the bushes or sneak into camp along the trail. It made me feel safer and I had decided that camp spot would be my home base. But the one we pulled into this time was

Figure 17 Open campsite

wide open and had the deep, dark forest opening up all around it. The sasquatch would be able to walk all the way into camp once darkness fell.

After everyone had retired on that first night, Cristy and I were sitting around the campfire catching up with each other. Cristy remarked about the number of frogs croaking out on the lake. They were really loud, and we started joking about what kinds of things they were saying to each other. That's one of the reasons I remember this conversation happening. We started cracking up and having such a good time sitting in front of the fire ad-libbing for the frogs. We retired shortly after that so we could get an early start at hiking the next day.

Cristy had stopped by the music store before leaving Portland and had picked up a new recorder for each of us so we could play some music together out in the woods. She had also picked up some cheap plastic recorders and pinwheels at the dollar store to leave out for the sasquatch. We were out to make new friends and what better way than through music?

After breakfast the next morning, we girls went for a walk while the men attacked the lake once again. While we were walking down the trail we passed a small creek. I looked down at it and spotted a small foot print. There was just a thin trickle of water coming down it at this time, and the print was left in the mud. It was not more than six inches long. We didn't have a tape measure with us to measure it accurately, so this is an estimate.

We then noticed that there was actually a trackway leading up this creek and we followed it up the hill. At one point the small tracks left the creek bed and then crossed a wet marshy area. As our subject crossed the moss and sorrel growing here, the tracks sank really deeply into the mud. At this point we were thinking that they could still be a human print. A human child could easily walk up the creek bed and into the moss and sorrel, and most kids love to walk barefoot through the mud. The trackway then continued on toward the tree line and the ground was getting a lot harder to walk on barefooted. There were needles, fir cones, rocks and tree branches littering the ground with blackberry vines winding their way through all of it.

I then spotted a new set of barefoot tracks coming out of the trees that joined with the smaller set. These tracks had to be at least sixteen inches long. The two walked back towards the marshy area and along the mud for awhile and then back in under the trees where we lost the trackway due to the hardness of the ground. We decided to go back down the hill to the lake. While we were walking back I asked everyone if they could destroy the tracks as we walked past them. I'm not sure why I felt that was an important thing to do, it just was and so that's what we did.

Cristy and I went hiking alone later in search of a good place to sit and play our recorders. She has such a beautiful voice and I love to sit and listen to her play her flute and sing. Since I didn't know how to play a musical instrument at all, she had brought the recorders to teach me how to play. We followed the road past the campground and walked up the hill. When we reached the top we had found the perfect place to play.

It was a little clearing with moss growing on the ground and forming a very soft carpet, with trees all around us and a ravine that fell away in front of us. We sat on the moss and tried to play a few duets, but I didn't know the notes by heart and I really wasn't getting the hang of it so I just sat and listened to Cristy play and then sing for a while. We were hearing the sounds of something moving around in the brush right below us on the side of the ravine the entire time we were there, although nothing ever crested the top of the ridge. We did walk to the edge and had a good look around trying to spot something down there but we never saw anything. Before we left we placed some granola bars in the branches of a tree and announced that they were for them, just in case we were being watched.

On our way down the mountain to the campsite we stopped at the creek to leave a few recorders lying on the ground. We left each one laying under one of the pinwheels that we stuck into the ground. We thought the pinwheels might catch their attention, and then they would see the recorders just in case they had heard us playing and wanted to try it for themselves. I had talked to a dear lady once who said that she had left some plastic recorders laying around her homestead and she later heard the sound of something trying to play one out in the woods and we wanted to see if the ones here would try it also.

The next morning we all talked about going down to the other lake so the guys could try their luck at fishing there. I knew I wasn't about to attempt to walk down there again, so we chose to drive instead. As we drove the road that led to this lake, I couldn't believe the amount of clear cutting that we saw. I could see at this point just why the sasquatch may have been upset about it. The entire back side of the mountain was being cut down, which meant that a good portion of their home was being destroyed. I do believe that would make me mad too!

When we got closer to the lake we started seeing huge logging equipment and loaders parked around the site, and I thought about Jerry Crew and his band of loggers in California who spotted foot prints around their equipment and even had some things thrown down into a ravine, and I wondered if these guys were having any troubles out here.

When we arrived at the entrance to the lake there was a small motor home parked there and a man came out to meet us. He told us that he worked for the logging company and that this lake was closed to unauthorized visitors for the season. We advised him that we were camped down below at the other lake, and he told us that it was actually closed too. He said the entire road was supposed to be closed because of the log trucks running up and down the mountain all day. Someone mentioned that we hadn't noticed any signs stating the area was closed, nor had we seen any log trucks on the way in, but we only came on the weekends when they had the day off anyway.

After joking with him for awhile, he said that the lakes were actually closed due to vandals that were coming in and messing with the company's machinery. At this point we were all just standing around talking in front of his motor home. He had left the door open when he had come out to greet us and I glanced over to it. I noticed that he had a very large gun leaning up against the dining table, and I remarked about the size of it. I remember him saying that he felt he needed it to protect himself from the wildlife around there. He said that he felt it had to be something big that was coming in at night and messing with the equipment.

At that point I looked at Cristy, and she looked at me and we both smiled at each other. He concluded by saying that he didn't have a problem with us being up there on the mountain, and we had permission to camp at the other lake also if we kept things quiet and didn't bother any company property. We assured him that we would keep it quiet and had no intentions of messing with company property, and also assured him that our campsites would always be clean when we left them. I then asked about my coming back up to camp alone in the future, and would I have permission to do so. He said it would be fine and told me to let him know when I was camping there and he would drive down occasionally to see if I was doing okay. I thought that was very sweet of him.

So the men went off in search of fish, and Cristy and I went off on a hike around the lake. The path that wound around this lake was very easy walking and was bordered by berry bushes. As we grazed our way around the lake, we came upon a small waterfall. We stood there enjoying it and I looked up and saw that we were standing right below the clear cut that my son and I had tried to cross, and I was kind of glad that I hadn't made it here the first time. It was definitely a hard hike getting down into here, and I could see that the climb back up would have been torture for me.

About dusk of that evening my husband started taking pictures of everyone in camp and shot a few pictures of the darkening forest around us. On one of the photos he captured one large whitish blue orb floating beside a tree, and next to the orb it looked as though something is peeking around the tree at us. Of course we saw none of this while we were sitting in front of the camp fire and visiting.

Figure 18 Orb outside of camp

As usual Cristy and I sat up watching the campfire after everyone else had retired for the night. We were just talking and having a great time. It was a very dark night with absolutely no moon shining and we were huddled up really close to the fire. The entire time we had been talking we could hear the frogs sounding off over the lake talking to each other, and I had pretty much tuned them out in order to hear her talking in a low voice. Then in mid-sentence she stopped and said, "Do you hear that?"

I said, "Hear what?"

She said, "Nothing. There is no noise at all. The frogs have quit."

I listened for a second and noticed that she was right; it had become deathly quiet out there. I don't know when the frogs had stopped croaking, but none of them was making a sound now. We sat and listened to absolutely nothing for a few minutes, and then I caught sight of her face, and she must have caught sight of mine because we both busted up laughing at the same time.

We sat there talking about what would have made all the frogs shut up at the same time like that. I mentioned that I had read an article that stated one of the signs to look for to know that sasquatch was in the area was that all of the insects and other animals would sense a predator and stop making noise. I reminded both of us that we had found the trackway going up the creek, and we had heard something down in the ravine while we were playing earlier, so who's to say they aren't around us right now?

I'm not going to lie to you here... we started to get a bit anxious. It was not from anything around us... we were scaring ourselves. We sat at the fire adding wood until we just couldn't keep our eyes open any longer. Neither of us wanted to leave the fire and walk to our beds. This was our first night in an area that we knew contained sasquatch and we didn't know what to expect. I had read stories of cannibal sasquatch, and woman stealing sasquatch, and even stories of friendly, helpful sasquatch. This could go either way, and we weren't sure which story to believe.

We finally decided that it was time to go seek our beds, or fall asleep at the fire. Cristy was going to climb into the back of the Jimmy, which was parked a good ways away from the firelight, and I was going to have to run for my tent, which was also a good ways away from the firelight and up under the trees. We decided that we would shine our flashlights on each other and walk backwards to our beds. That way we would be able to see something coming up behind either of us, and be witness to any abduction that might take place.

We knew we wouldn't have a chance at reaching the other should this actually happen, but at least we would be able to tell other people what happened and have a witness to the event. I thought that it was a great idea, especially since I was too scared right then to walk her to the truck and then walk back to my tent alone. So this is what we did, and when we got to where we needed to be we counted to three, and I quickly unzipped the tent door while she opened the car door and we both jumped into what we felt was safety.

I quickly got into my sleeping bag and fell asleep. The next morning Cristy told me that she had awakened at some point and heard something next to the truck. She said that she had looked up and saw a shadow standing towards the back of the Jimmy through the side window where she was laying. She said she closed her eyes immediately and never looked up at the window again. I believe her story and it has never changed. I didn't know enough at that time to analyze the truck for hairs or smudges on the glass.

We just got a good laugh at how brave she was. But in all seriousness, I probably would have done the same thing.

The next day Cristy and I planned to go on another hike together and I took her off into the forest on the same path I had followed the first day I had set foot in this area. I wanted to be able to follow the creek all the way to the bottom and this time I had someone else with me. The first time I had followed this path, I had become scared and turned around and went back to camp. I figured this couldn't happen a second time, because now I had someone else with me who could keep me lucid.

We were having a wonderful time slowly meandering our way through the forest and looking at all the beautiful scenery that we were walking past. We actually passed the spot where I had gotten scared the last time and I didn't say a word to Cristy as we passed it by and just kept walking. We eventually got to another curve in the trail and this time Cristy stopped dead in her tracks. I stopped next to her and asked her why she stopped, since I didn't really see anything laying around that looked interesting enough to stop and check out.

She looked straight ahead and shook her head slightly and said that she didn't think she was supposed to go into that area. She didn't feel like she should be there. I understood the feeling very well, but tried to get her to agree to move forward anyway. She wasn't having any of it and I totally understood, so we turned and continued hiking into a different part of the forest. I don't know what it is out there that could keep us out of this area, but I can honestly say that up to this day I have never made it all the way down that hill.

There wasn't any activity outside of camp that we noticed that night and we just spent a pleasant evening around the campfire making Smore's and having a good time.

Chapter 9 – Late Night Visit - Orbs

The following weekend, I returned to the mountain to camp with the people who had originally shown us the spot and, for some reason, they retired really early that night. It was only them and me on this trip as my husband and son had stayed home. It was only nine pm when they zipped up their tent. I briefly thought about taking a hike but remembered that there was a large group of campers across the lake. I really didn't feel like running into anyone and having to strike up a conversation. I decided to just crawl into my tent and sit on my bag and do a little reading instead. Right around the time it was getting dark I heard a car pull up and park on the road above camp. We weren't expecting anyone to join us up there, so I peeked out of the tent and looked up to see who it was. It was a couple that were walking over to the lake instead of coming down into camp. It was getting too dark to read by then, so I turned on my flashlight so I could see better and thought no more about them.

A short while later I began to notice that my tent was all lit up on the inside and I was able to see my book without the aid of my flashlight. Once I took my concentration off of what I was reading, I also began to hear the couple at the lake talking about an Unidentified Flying Object (UFO). I started listening to what they were saying and they mentioned the UFO again, and were saying something about how it may have landed on the mountainside across from us.

As soon as I heard this I realized that that must be why my tent was all lit up. I slid off my sleeping bag and unzipped the tent and crawled out looking up to the sky to see if I could spot the UFO. There wasn't anything going on in the sky, but the entire mountainside across the lake was lit up like the mother ship had just landed there. I kid you not!

I walked down to the lake and stood near this couple who stated that they didn't know what was going on up there. The forest was lit up like it was broad daylight and we could see each individual tree and fern illuminated on the ground. I stood there with my mouth hanging open for a couple of minutes just wondering what the heck was going on. I was half hoping that it was a UFO, and half hoping that it wasn't. On the one hand I couldn't deny that they were real if I saw one with my own two eyes, and on the other hand if I did see one with my own two eyes my world would never be the same.

As we stood watching, the lights all started to separate from each other, and move off in opposite directions. We heard a man's voice call out across the lake through a crackly walkie talkie and I noted then that the lights must be individual spotlights, and they were starting to spread out around the mountainside. There were people walking all over the place with mega flashlights and night vision! They had to be here searching for the sasquatch. I had read about these million man expeditions, and how disruptive they were for the sasquatch families that lived in the area, and I began to get so incredibly mad at that for some reason.

The couple I was standing with asked me if I could figure out just what they were looking for over there, and I feigned ignorance on the subject. I didn't want it broadcast that the sasquatch were living there and I just stood in that spot fuming for a few minutes. I was so ready to hike across that lake and start slapping the heck out of everybody over there. As I said before, I was getting really protective of this sasquatch family already, and I felt about as violated as they must have felt. I told myself that the best thing to do was just to let it go for now, and I would go over and have a chat with these people tomorrow and find out exactly who they were, and what they were looking for.

I walked back into camp and sat down in front of the campfire thinking about what I would say to those people the next day. The other couple never once woke up to see what was going on, and I never heard their two pit bulls rustle either. I decided I would just go ahead and cook the hot dogs I planned to eat for lunch the next day, and maybe eat a few. I got out the metal hot dog skewers and a pack of hot dogs and settled down in front of the fire. I put the entire package of hot dogs on four metal skewers and then leaned them up against the fire pit so they were hanging above the fire. I sat back and thought about things for a while. Like why the couple that I came with went to sleep so early? And why haven't they woken up with all the yelling echoing back to us over the lake? Why haven't the two dogs begun barking, or even moved about in the tent with all the commotion going on around them? Who those people were across the lake trekking all around the mountainside? Didn't they know that the sasquatch came to you? ...Things like that. But, what was bothering me the most was why were the dogs being so quiet? They weren't like that at home.

As I sat there contemplating these things, I heard a soft whistle behind my right shoulder. It sounded like it came from directly behind me. I was sitting with my back to the bushes that encircled the camp and I then noticed that I was sitting so close to the bushes that I could feel a branch poking into my back. Now I generally don't like to sit with my back to those bushes, and I make it a habit not to sit this close, but tonight I had been so deep in thought that I had just sat down on the big stump that we usually used as a table next to the fire pit. It was sitting right next to the fire ring and we had placed the chairs across from it out in the open where I usually sat. In case one wonders why I don't sit here, this is the area where the sasquatch make their appearance and, at that time, I wasn't comfortable enough out here to turn my back on them.

When I heard the first whistle, I did nothing... I just sat there wondering what would happen next. I don't know why it didn't concern me, or scare me even and I still find it surprising to this day how calm I actually was. After sitting for a few minutes, I heard the same soft whistle come from behind my left shoulder. I had never heard any rustling from the bushes to let me know it was moving, but it had to have shifted at least slightly for the sound to have changed sides.

At this point I knew it was hungry. I can't tell how I knew this... all will just have to take my word for it. I simply KNEW that it would like to have something to eat. I actually chuckled out loud right then too because the second thought that came to my mind was that I was sitting here angry at the fools across the lake that were looking for the sasquatch and this sasquatch was smart enough to just walk over to this side of the lake and hit me up for some good grub. I figured it wanted to sit and have dinner and watch the show happening on the other side of the lake. That put me into a much better frame of mind and I started smiling.

I quietly said, "Are you hungry? I'll put some dogs into some buns for you." I continued sitting where I was and reached over and grabbed the package of hot dog buns from inside the dry cooler. I picked out three hot dogs that looked warm and placed them inside three buns. I started reaching for the condiments and then realized that I didn't know if it would like mayo, ketchup or mustard on its hot dogs and actually thought about asking! This made me laugh out loud and I decided it would be best just to leave them dry.

I hadn't heard another sound from it in all this time, but I didn't let that put doubt into my mind. I placed the dogs on a flimsy paper plate to serve them. I figured that I would put the plate up on the cement pole that stood right outside of camp next to the incoming trail. I think that this pole used to be where they would chain a garbage can at one point in time. It was about two inches smaller in diameter than the plate. I would have to balance the plate on top of it. I figured this way I could also be sure that something with hands was eating it because if a bear or a coyote or something smaller tried to get to it they would knock the plate off onto the ground reaching for the food.

I don't know where I found the bravery to walk out of the firelight and towards this thing in the bushes but as I said I wasn't scared at the time. Not the least bit. I walked over to the pole and balanced the plate on top of it. As I was doing this, I was talking and saying, "These are for you to eat, I have more." When I was done balancing them on the pole, I took a few steps back away from the pole and said, "You can just reach your arm through the bushes and get them and no one will watch you. I'm going to go into my tent to let you eat in peace."

I walked back over to the fire and got the other dogs off the skewers and put them into the cooler. I banked the fire and went over and got into my tent. The sky was still lit up and I could still see well enough in my tent to read without a flashlight. That's how bright the area was from all the lights on the mountainside. It really didn't bother me anymore because I knew where at least one of the sasquatch was hanging out... it was safe and sound and eating dinner. After I had read a few chapters, I decided that I wanted to go see if it had actually eaten the hot dogs or if I was just making stuff up in my head.

When I reached the pole I noticed that the plate was still balanced on the pole, but the hot dogs were gone! I never heard the other couple unzip their tent and I was in seventh heaven knowing that the sasquatch did indeed get to sit and eat dinner. I did find a small piece of the end of one of the buns on the ground next to the pole. It showed no teeth marks or other holes in it. It appeared as if someone had torn off a small piece as opposed to biting it off. I left that piece lying on the ground so the baby opossum that had been in camp earlier could eat it. I said "Good Night" and floated back to my tent. I was so happy that night, and I decided then and there that there would no longer be any doubt in my mind. The sasquatch existed, and no one could tell me otherwise.

The next morning I asked the other couple if they had heard anything the night before, or noticed how the lake and the surrounding area had been lit up like it was daytime from around eleven or so until around three am. They both said that they had never heard anything and the dogs hadn't awakened them either. I told them about the other campers across the lake walking around the mountainside with flashlights and night vision, but never told them about my visitor.

I thought about going over to the campers across the lake and asking them some questions about what they were doing up there, but I didn't want to give them too much information about myself at the same time. I had no desire to be known to the sasquatch research community and I knew that they hadn't found anything over there. The sasquatch had proven to me that they were not dumb apes. They had intelligence, and they knew how to outsmart us. I did hike around the lake the next day to get a look at their campsite though. The thing that surprised me the most was the three little children running around camp. They had to be between two and five years of age. My first thought was "Bigfoot bait? Really?"

Let me take the time to explain that thought. I was very new at this and I really knew nothing about this species except that, judging by the size of their tracks, they must be extremely large and powerful. I'd been reading different theories that stated the sasquatch love watching women, and children at play. Some web pages even suggested that we take our children camping with us to force the sasquatch to come out of hiding and check out the kids. Seemed kind of unsafe to me when I read that, especially if we don't know if the sasquatch that are coming around are friendly or not. Plus at that time, I was still stuck on the negative perspectives that I had been reading.

I walked by this camp wondering exactly why they would bring their children to camp knowing that they were out there to 'rustle up' some sasquatch. I took a quick survey of what I could see laying around and kept walking up to the ridge above them. I wanted to look around and see if I could find any footprints or places where a cast may have been pulled up out of the ground. I found a lot of human boot prints, but no sasquatch prints or proof that any had been found.

I hiked around the area most of that day and stopped to try to play a few songs on my recorder in the same clearing that Cristy and I had played in before. I still wasn't too good at it, but I had fun trying. I hiked back to camp and found that dinner was ready and I was very grateful to see a hot meal prepared and waiting for me. After dinner the couple once again retired to their tent while the sun was still up. I sat in front of the fire until the sun had just touched the top of the mountain and the day began to wane.

I had brought along the night vision that Oldie One had sent to me and wanted to play with it after night fall to see what it could do. I wandered down to the lake to play with the night vision down there. I walked out on a large log that had one end pulled up on the bank and the other end floating on the water, and I sat down at the end of it. I was watching the fish and salamanders swimming in the water while I waited for full dark.

This is a round lake with two docks, one on each side of the lake, and a lot of snags in the water. Where I was located on the water I had one dock to my right, and one dock straight ahead of me. I loved sitting over the water watching the sun set from the end of this log. The sun reached the horizon and the trees turned black against the still lit sky. The air under the forest turned an indescribable color of gold, and a hush fell over the forest. The diurnal creatures go to bed, and the nocturnal animals begin to stir. It's just such an amazing time to be in this forest.

As I was sitting there feeling peaceful and watching the sky change to black, I figured it had to be dark enough for me to try out the night vision. I pulled it out of its bag and dialed it in correctly and started to scan the shoreline around the lake. I was having so much fun with it and was quite surprised at how well I was able to see everything, albeit I was seeing everything in green. I started scanning from my left side and circled around the shoreline until I came to the dock directly across from me. I expected to just see an empty dock and pass on by it to continue scanning the shoreline, but instead my eyes focused on a guy who was standing on the dock with a red light flashing between his eyes. I looked closer and saw it was a man standing there holding night vision to his eyes and he was looking right at me!

I knew that he could see me over here, since I was seeing him over there, and I raised my free arm ever so slightly and waved at him. He actually raised his free arm and waved back! That was kind of freaky. I felt like I had been caught peeking into someone's window, so I pulled the night vision away from my eyes. I looked over to the dock with my eyes and could just barely see a shadow shape standing over there. I remember thinking how much I loved having the night vision and how well you can see in the dark with them.

I placed the night vision down on the log next to me and figured that I would use them again after he had left. I had all night to sit here and now that I knew how well the night vision worked I planned on playing with it some more. I thought about walking through the woods with them, or looking down the road to see if I could catch a glimpse of a sasquatch, but I was enjoying myself just sitting there looking out over the water. I then noticed the guy on the docks put down his night vision and turned and started off down the dock towards where he and his friends were camping. I could hear his boot heels hitting on the wood in the still night air as he began to walk down the dock. About the time he started walking, something totally strange happened… Another Twilight Zone episode for sure.

As I sat there looking towards this dock, I noticed a light out of the corner of my left eye and I turned my head in that direction. I was wondering how someone could sneak up on me when they would have to walk across gravel to get to that spot. I thought it was a flashlight coming out of the trees, and I hadn't heard a thing. I was looking at it, but I couldn't believe what I was seeing with my own two eyes. It was a round shining ball flying through the air!

It was larger than a softball, but smaller than a basketball, and it was shining with a very bright bluish/white light but more white than blue. It wasn't pulsing, or glowing, it was just shining with this wonderful light. It was so bright and yet it cast no light on the vegetation around it. Nor did it cast a light on the water when it crossed the lake. As it continued to fly out of the trees and then through the bushes that encircle the shoreline, I noticed that it also had a tail attached to it. The tail was a part of the whole and not a separate piece, so the entire object was shining as one unit. It wasn't transparent, and I couldn't see through it. It was completely solid throughout the length of it.

It was flying out of the tree line when I first spotted it, and it continued to fly in a direct course over the top of the lake about six feet off the water. It had to be that high to miss the many snags standing in this lake. It never rose up and it never fell down. It just flew in a straight line across the lake. To my mind that instantly left out the possibility that someone had thrown something out over the water. So obviously, my mind was trying its hardest to make sense of what I was now seeing.

The closest thing I could find in my mind's data bank to describe this thing was a shooting asteroid from those Glow in the Dark sets. I had the Universe set plastered all over the walls of my bedroom years ago, and it came with stars, planets, and astronauts. One of the pieces was an asteroid with a burning tail attached to it, and that is exactly what this looked like. On a much larger scale of course, and the tail wasn't on fire, but it resembled this so closely that the first thing I said in my mind was, "Oh look, a shooting star!" I quickly reminded myself that shooting stars are in the sky, and this was over a lake. "Well, then it's a shooting comet!" ...Over a lake?

As I was sitting there with my mind futilely trying to figure out exactly what this thing was, I was still watching it fly over the lake and it was now directly in front of me. It was very solid and hadn't quit shining the light. The light seemed to be a part of it, as opposed to just being a light. It seemed to be the light, and the light just is. Like it's definitely an organic kind of thing and not just a metal or mineral kind of thing with a light in it. So now I was just telling myself that, "Hey, it's an orb, and I'm cool with that."

I had been picking up orbs in most of the pictures we were taking in this area. Although, it was really bothering me that I was seeing this thing with my own eyes and not just in a picture. If an orb shows up in a picture we can tell ourselves that the lens was dirty, or the lens had water drops on it. But when you are sitting and seeing one right in front of you, all pretenses are off and there is no denying what you're seeing.

I watched this orb fly across the lake and I noticed that it was headed in the direction of the dock that the man was just standing on. It wasn't moving fast, it was just leisurely making its way across the lake in front of me. I was wracking my brain the entire time trying to make sense of the sight and I must have been too enraptured in the moment to feel frightened at the time. I got a really good look at it while it was flying and I will never forget what I saw, yet I still have a hard time putting a description of it into words. It was there, it was solid and it was light. I saw no seams of any kind and no edges of any kind. It was just a solid light ball with a tail that never wavered in its path.

So I've watched this thing fly across the entire lake now and it's was getting really close to that dock. I was thinking it's going to hit the guy as he's walking with his back to it, or maybe crash into their campsite. I was about to yell across the lake to alert them in case they hadn't seen it when the darn thing just disappeared! My mind went into a mild state of shock just then and my brain was screaming at me, "Where did it go?"

"The light went out! They turned the light out!"

"No, they didn't turn the light out, it vanished!"

"Vanished?!? It disappeared... It disappeared??"

And it had... right in front of my eyes! I'm not lying, it was there one minute shining brightly, and the next second it just went out, vanished and or disappeared. It was still about six feet off the lake at the time it went away.

I sat there straining my ears to see if I could hear it crash into the woods, or onto the dock, but I didn't hear any sounds of any kind. I know that it never fell into the water, because I never heard a splash. I also listened to see if I could hear this other guy or his friends saying anything about it, or getting hit by it. I heard no sounds at all... just total silence. So it didn't hit the guy, and he wasn't screaming about anything, and the woods didn't burst into flames and nothing splashed into the water. I had no idea where this thing went. Heck, I still had no idea what this thing was, so after sitting in the same spot for another few minutes and listening to absolutely nothing, I gave up and retreated.

I said out loud, "Oh heck no! Okay, that's enough for me for one night; I'm going back to camp!"

I stood up and walked off the log and made my way back to my tent to grab my journal and get all this written down.

Of course I didn't have a camera with me at the time, and that doesn't surprise me in the least since that seems to be the times I have the best experiences. I had to get that image out of my head. When I got home I grabbed a picture of the lake that I had previously taken from that same exact spot, but during the day, and used Paint to draw a reproduction of what I saw onto it.

After walking back to our campsite I sat in the open doorway of my tent. I was just sitting there as still as I could and listening for any sounds and wondering if I would ever see the Orb again that night. No one started screaming and running for their cars and I never heard a sound. I was having a really hard time wrapping my head around what I had just witnessed. My brain really had no prior knowledge of anything like this to access so it was busy scanning all the stored files on all the things I had ever seen, or read about, or heard about or maybe caught a whisper about, to try and describe what I had just witnessed. My mind wasn't having any luck, so I finally decided that I would just have to go with the saying, "it was what it was."

I couldn't let it drive me crazy... I would just have to go with it. That's what I had been doing up to this point anyway... Just going with it, and letting what was going to happen, happen. What else could I do? This was a very magical spot, and magical things just seemed to happen here. Isn't that exactly why I kept coming back here anyway? I let it go for the night, but I had one final thought as I climbed into my bag and fell asleep with a smile on my face, "How many people can say that they have watched a glowing Orb lazily fly over the top of a lake?" I know that I am very thankful to have this experience to look back on.

Figure 19 Orb painted onto picture of lake

Chapter 10 - Vision of Me

I wanted to return the next weekend but found that no one else was available to come with me. I loved this spot and couldn't get enough of it, so I decided that I would just do it by myself. The only thing that I wasn't completely sure of was just exactly which of the many logging roads I was supposed to turn on to get here off of the main road which led up the mountain. It seemed they all looked the same here on this mountain, and none of them were marked anymore. I figured I could just have to drive every one of them to find it if I had to. I was also secretly hoping for another visit from my new forest friends, but would never have dreamed of the experience I was about to have.

I drove up the mountain alone and started to look for the proper road to follow. I was beginning to wonder if I would find the correct one before dark when I spotted a very large rock stack sitting on the side of the road at one of the junctions. It was really quite large and I had never seen a rock stack like it before. I pulled the Jimmy over and sat looking at it for the longest time. I had a strong urge to draw it and get it into my journal, so I did. It was comprised of a very large flat, round rock on the bottom, and two stacks of three square rocks which leaned in toward each other and met at the top. As I sat looking at it, it seemed to scream two in one to me. This made complete sense to me as I had been told that there were two different clans who shared this area, and their homes met at the lakes.

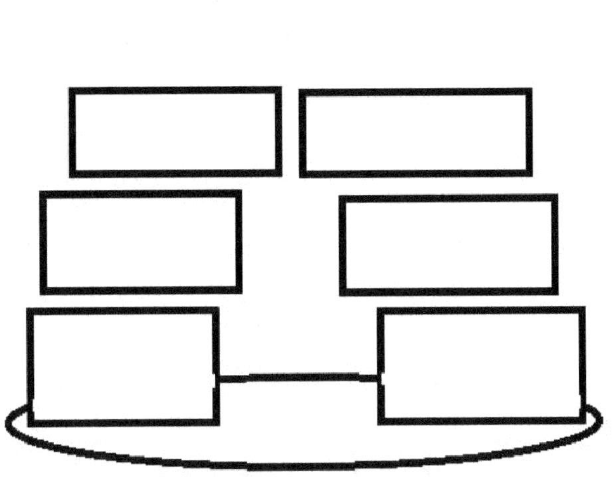

Figure 20 Rock Stack

After I made the sketch, I turned down this road and found that it was indeed the road that I had been searching for and it led me right to the lake. I went straight to the secluded campsite, the one that is surrounded by berry bushes, and set up my camp. I was so happy to see that I was the only person up here so far and I hoped to be able to have the place to myself for the weekend. I spent the morning doing a little hiking and spent the afternoon around the lake.

When sunset was approaching, I went back to camp and built up the campfire and prepared something to eat. I was sitting down in my camp chair with my back toward the tent and the fire in front of me a short ways away. As the sun descended in the west, the forest turned a beautiful golden color and I sat enjoying the show until the sun had set below the horizon. I then began to practice some words I had heard on an internet radio show. They were supposed to be sasquatch words that a lady was sharing with everyone. I had copied them down in my notebook and was now sitting in my chair with it open on my lap. I was holding a flashlight in one hand and had it shining down on the words so that I could see them. I was repeating the words out loud in as much as I could remember how to pronounce them. I'm sure I messed quite a few of them up by pronouncing them wrong. But I was eager to learn how to communicate with the sasquatch the next time they came into camp.

As I sat there reading, and mispronouncing these words, I suddenly became aware that I now had a different vision in my eyes. Instead of just seeing the page that was sitting in my lap with a circle of light on it and the bottom edge of the flashlight, which was my normal vision, I could now see the top of my head. And I could see my hand holding the flashlight all the way up to my elbow. I saw the notepad sitting on my lap, but I could no longer make out the writing on the page. I'm not exactly sure at what point my field of vision changed, but as I began to notice these changes I saw that my head appeared to be getting farther away from me and getting lower and lower.

I started to feel a bit queasy inside because it was just like I was shrinking, or getting smaller, or something! But it was only from a visual standpoint. I myself didn't feel any change at all. This really panicked me because it actually looked like I was looking at myself, while behind myself, while I was quickly standing up. It didn't make any sense to me at the time and it's still a very hard thing to describe, so I hope the understanding is clear in what I mean here.

That's when I started to get the feeling that something must be standing behind me and somehow it had projected a vision into my head of myself sitting there... Either that or it was actually able to allow me to see through its eyes! Either way I was disturbed over all the strange things this species was doing around me and I got so scared that I actually jumped up out of my chair and yelled, "Just don't hurt me" really loudly.

I jogged toward the fire and never once turned around to see if anything was actually standing behind me. I really don't know if I could have handled it had I turned around right then to see something eight to nine feet tall standing there. I began to speak out loud and said, "Oh no you don't! I don't think I can handle this! If you want to see me you are more than welcome in my camp, but come straight at me. None of this sneaking in behind my back stuff!"

By now I am standing in front of the campfire and I still haven't turned around. I continued speaking and said, "And none of this 'counting coup' I keep hearing about either! Don't run up behind me and try to tag me. I want Good Will. I love and respect you, and I ask the same in return from you. Don't try to scare me anymore, just come on into camp and say hello."

After my tantrum was over, I stood by the fire for a few seconds and actually burst out laughing at myself. I was wondering what the other people around the lake, if there were any, must be thinking about me suddenly yelling, "Just don't hurt me" out in the middle of the forest like that. It had been so quiet, and I had said it so loudly that I could hear it echo back at me as I was walking over to the fire. I stood there looking out at the trail that led to the forest for a while, but no one ever did come into camp to say hello.

At one point, however, I thought I saw a pair of glowing eyes on the other side of the bushes in the campfire light. I stood there and tried not to look at it for so long that it finally started unnerving me and I scared myself enough that I just had to shine the flashlight in that direction. I had been told to never shine a white light in their direction or the fun is over for the night but I finally had to see just what it was that I thought I was seeing. That night I learned how easy it is to scare oneself out in the woods because it turned out to be a couple of Salmon Berries glowing in the fire light. This made me realize just how much I wasn't ready to come face to face with a sasquatch while I was alone.

I had also been told that if I set a soft light down a few feet away from the tent, I would be able to see their shadows cast on the tent if they came into camp. Before I retired for the night I turned on the battery operated lantern I had brought with me and set it down by the bushes. Before I could even get back to my tent the darn thing went out! I walked back over to it and picked it up to check it out. Everything looked good, so I turned the switch and it came back on. I set it down and started to walk back to my tent and it went off again! I tried to get it to come back on again but it just didn't seem to have enough power, so I brought it back to my tent with me.

I then said Good Night to the forest and zipped up the tent. I tried to get on my computer and the darn battery died even before it had booted all the way up. I figured that the Universe must be trying to tell me something and it must be time to just go to sleep and call it a night. Of course, I stayed awake for a little while listening for any sounds but I never heard anything and finally fell asleep.

This camping trip really upset me and, since all my batteries had died the night before, I decided to just head back down the mountain the next morning. I did take the time to write everything down in my journal before packing up camp. I knew I had to write it all down while it was still fresh in my mind and I wanted to be able to remember every detail later. I knew I was going to have to research this event and see if I could find others who had this type of thing happen to them. To this day, I have heard of no other event like this one taking place, and I am still interested in meeting anyone else who has had it happen to them.

My husband and I camped here again the weekend of June twenty-sixth with another couple but I had no other experiences that weekend. It seemed nothing much happened when the others were around. I was kind of glad because I was actually getting tired of the Twilight Zone stuff and I just wanted to enjoy myself this time without any mind altering events happening that I would have to try to wrap my head around later.

I had already come to grips with hearing voices, and seeing visions, but I was still having trouble with seeing myself from outside myself. And what about them knowing what that guy was up to when he was walking up the road? I guess they could have seen him walking up the road and figured it all out. But that would clearly take them out of the animal realm now wouldn't it? And why would they come to protect me if they were cannibals, or woman stealers? I had a whole lot to think about already and a nice normal weekend was a plus.

Cristy was coming over for the Fourth of July weekend, and we planned on heading up the mountain to camp again. Everyone for miles around would be in town enjoying the holiday festivities, and we were hoping to be the only ones up on the mountain that weekend. When she arrived at my place, we loaded the things that we would need into the truck and left town with quickness. When we arrived at the lake we were jazzed to see that no one else was camping up there and we set up camp with smiles on our faces... Then the couple who showed us how to get here originally arrived. They drove up and wanted to make camp with us in our campsite. Neither Cristy nor I wanted them to be there, but we decided it would be best to be hospitable and just let them stay.

Once again they had both of their Pit Bulls with them. Remember that this was the couple who had been with us when my husband and I found and cast the foot print in the mud and this fellow had given me such a hard time about it. He brought up the sasquatch again as we were all sitting in front of the fire, but he had a much better attitude about it this time. He actually seemed interested in the subject and I shared a few of the reports I had recently read on the internet and we had a great time talking about them and what they may be. Of course, I shared none of the experiences that I had recently had in that same campsite. Those were my own, and I wanted to keep them private.

The next morning, Cristy and I headed out alone to explore more of the area and we decided to follow the little creek up the hill and see what we could find up there. When we got to the top we discovered a gravel covered spur road that ran from north to south in front of us. We walked out on the road, and then followed it uphill until we came to a T intersection. The road to the left of this intersection continued on uphill, and we were getting pretty tired of walking uphill at that point, so we decided to take the flatter road to the right.

We started seeing beautiful wildflowers blooming in the middle of the road and they were getting thicker as we continued along, until they started to take over the road. That wouldn't have been a bad thing, but along with the flowers came the bees. We figured we didn't want to test our luck walking there without getting stung, so we'd best find a way to skirt the road through the trees instead. I'm so glad we did because we found the most perfect little clearing in there. We spotted a huge fallen cedar tree and made our way towards it. We got comfortable on top of it and Cristy pulled out her flute and started playing and singing hymns. I just sat back and listened to the beautiful sounds floating on the air of the forest.

After playing for a short time, Cristy asked me if I felt like I was being watched. I had been sensing that for a while but never mentioned it to her. When she asked me that question I realized that it wasn't just me with that feeling and I started looking around and wondering where they were right then and if they were enjoying the music as much as I was.

We eventually decided to wander on a bit further and see what else we could find. As we were passing a large tree, I glanced down towards the roots and noticed a rather odd thing. There was a long bone lying on the ground. It had been tucked under a piece of root that came up out of the ground and then went back in to the ground forming an arch. The bone wasn't buried in the dirt, and it wasn't even dirty. I reached down to check this out and Cristy remarked on how it seemed that someone had put that bone there under that root, and maybe we should just leave it alone. As we were leaving the bone, I looked up and noticed a worn path leading into some bushes so we decided to walk over and see where it led.

We followed this path through the bushes and found ourselves on the edge of a huge clearing with the sun streaming down on us. It was a beautiful place with high grasses and lots of downed trees and huge cedar stumps in the middle of it. We both just stood there staring into this clearing, but I couldn't seem to get myself to step out into it. The air was too still here and it just seemed to give us caution. Cristy said that it looked like the perfect place to lie in the grass and sun oneself and I agreed wholeheartedly. We discussed walking out into the clearing and sitting in the sun for a few minutes, yet we still just stood there looking and never moved. I think we both decided at the same time that this was not our spot. This was somebody else's spot and we just couldn't seem to make ourselves invade it, so we turned and walked around this clearing instead of entering it.

We eventually found ourselves at the beaver pond that my husband and I had found previously. We sat on the bank in the sunshine and waited for brother beaver to make an appearance, but he never did. We knew that we were not alone however. The only sounds we heard were a slight rustling up on the hill above us moving off to our left. I turned to look that way, but of course I never saw anything. Cristy pulled her flute out and began to play, and it was so nice and relaxing that I could have sat there for the rest of my life just enjoying the good company and the wonderful music but eventually we had to return to camp and leave this beautiful spot for another day.

About an hour before dark everyone agreed that we would go down into town to check out the fireworks show. We decided to drive down in separate cars so that we wouldn't have to wait for each other when it was over. It's a good thing we did because Cristy wasn't having any fun and wanted to head back up the mountain almost as soon as we arrived. When we pulled into the campsite it was pitch black there and we could see absolutely nothing out of the truck windows. I couldn't even see the bushes where we had just parked. They were so close they were touching the trucks bumper... That's how dark it was out there.

Neither one of us was in a big hurry to get out of the truck at that point so we sat there laughing about what brave princess warriors we were. We always call each other that, and it usually gives me enough incentive to get over any fear, but I didn't want to have to walk into a pitch dark camp and have to try to find the lantern by myself that night. I knew what was happening there and I didn't relish the thought of running into camp and running right into one of them. We decided to leave the headlights on and light up the area until we got the fire stoked, and believe me, we stoked that fire up in a hurry. We then got comfortable and sat around the fire talking and laughing about our bravery, or lack of it, and enjoying the quiet night until two am when we decided it was time to get into the tent. I know the time for a fact because from the very start I have made it a rule to get into my tent at two am and then sit and listen for them to come around camp.

The other couple arrived right about that time and the wife went straight into the tent with both dogs. I could hear the man doing something around the camp fire area but that area was blocked by my tent so I couldn't see exactly what he was up to. We had placed my tent so the door was looking out towards the lake instead of in towards the camp site. Cristy was lying in her sleeping bag and I was sitting on my bag in front of the open doorway just looking out into the darkness towards the lake. I saw him walk past my tent and I froze instantly. I didn't want him to know I was still up, and I was hoping that he hadn't noticed that my tent door was open.

He walked around the corner of the bushes and headed towards the lake, which was just across a one lane dirt road. All of a sudden I saw bottle rockets shooting out over the lake. It was really pretty and I turned to Cristy and told her to get up and look at the pretty colors over the lake. She said she was sleeping, and didn't want to wake up so I let her be and continued watching the fireworks.

Just as the last rocket burst over the lake, I heard the loudest, sharpest tree knock that I have ever heard. Just one huge THWAK! Just one time, and it sounded like it came from directly behind this man! Now it must be understood that everyone in our group knew for a fact that we were the only humans at this small lake. My eyes flew open and I knew they were angry! The strength that was put into this tree knock left no doubt in my mind.

Where this campsite meets the road there are three steps that were cut into the hillside, and two by fours were placed across them to form steps. After the tree knock I heard feet slapping the dirt road at an amazing rate of speed and coming towards the campsite. I saw the husband as soon as he hit the top of the steps. His eyes were huge, and I swear the fear in them was like nothing I had ever seen before. He flew into a slide that would have been the envy of every baseball player on earth. He slid right down each one of those steps and was aiming right for my tent, all the while blabbering, "They're here...They're here...They're here…Help me!"

I just stuck my right foot out of the tent and kicked him just hard enough to turn his slide off to the right, and he slid past my tent blabbering, "You were right…They're here!"

I called out to him as he slid past and told him, "You angered them, you deal with them!"

He got to his feet and stood there staring at me like I was going to let him into my tent or something. I told him to make a run for his tent, and he turned and ran as fast as he could. As he unzipped his tent door I thought for sure the dogs would be barking or at least awake and want out after what had just happened, but they weren't. Even the wife slept through it.

That's when I remembered Cristy lying next to me in the tent. I leaned over and whispered, "Wake up" and "Did you hear that?"

She wouldn't wake up. She just said, "I don't want to wake up! I'm sleeping! They won't hurt me if I'm asleep!"

I figured I couldn't argue with that logic, and so I just sat there quietly with the door open for about twenty minutes. I was listening to the sounds of the night, or should I say the lack of sounds of the night. I kept trying to get Cristy to get up so we could go out and see what was up but she was having none of it. I wasn't going out alone, so I finally zipped up the tent and went to sleep.

The next morning we all talked about what had happened the night before and laughed about the perfect baseball player impersonation. His wife stated that she hadn't heard a thing. We all wondered why the dogs hadn't awakened either, even when the rockets were whistling out over the lake.

Later that afternoon the husband took the dogs for a walk, and came back into camp with multiple bee stings on his body and beginning to puff up. Even the dogs had gotten stung while on their walk. He said that they must have stepped on a bee's nest because suddenly he and the dogs were all being swarmed by bees. Since we had planned to leave that day anyway, they went ahead and packed up and left.

I was thinking it was just an accident brought on by being in the wrong place at the wrong time, but after they left Cristy reminded me about how often we had walked around in the exact same location and nothing had happen to us at all. She was also wondering if the sasquatch could make the bees attack him because of what he had done the night before. I really didn't know if they could do that or not, but it seemed he had made them quite mad. Plus he was always putting them down out here... saying that they didn't exist and making fun of us for believing. I don't know if they can do this sort of thing or not, but wouldn't it be an excellent payback?

I returned alone in August to get some much needed forest time. My home life was falling apart and I needed the solitude and peace to get my head straight. I drove up the mountain not knowing what to expect, I just knew that I needed to be there. I was delighted to find that the campsites were all deserted and I was alone at the lake. I set my camp up in the isolated campsite and hoped that no one else showed up that week.

I walked out to the lake to watch the sunset and do some thinking after which I walked back to camp and started a fire. While I was doing this, I heard cars drive past me and head to the back side of the lake. I was hoping that they hadn't seen my fire from the road and hadn't noticed that I was down here.

It was fully dark when I started hearing very loud music coming from their campsite. I didn't mind that so much because they were playing some really good music and I found myself dancing around the campfire. I felt so free being out there alone and just having fun by myself that I lost my inhibitions and let the music move me in whatever way it wanted to.

I ate dinner around eleven pm and sat in my chair in front of the fire for about an hour, still listening to the music coming from the other campsite. When I heard the music shut off and everyone saying goodnight to each other, I banked the fire and crawled into my tent. I took the ax into the tent with me and laid it on the floor just in case I needed a weapon later. I also had my wood knocking stick with me and I placed that on the floor of the tent next to my sleeping bag.

I crawled into my bag to get comfortable and started hearing the sound of something walking along the road above my camp. I was situated below the road, and it sounded like whoever was up there was coming my way.

I lay there listening to these footsteps walking along the road and getting closer to my camp. I wanted to listen to if they would stop above me, or if they would just continue on down the road. That was when I heard the distinct sound of three heavy footsteps coming across the gravel from the backside of my camp. It walked right up to my tent and stopped… Right next to my where my head was laying.

This concerned me deeply and I wondered if the other campers were trying to circle me and do me harm. I reached over and grabbed my walking stick with the intent to use it on someone. As soon as my hand wrapped around that stick I was bombarded by a feeling of concern and uncertainty.

I then realized that whatever had walked up to my tent had only taken three steps into camp, but had covered a great distance of space in those three steps. I still had my hand on my stick and was planning how to best use it to subdue whoever it was that wanted to do me harm.

At that point, I heard a quiet voice ask me, "Why do you want to do that." I now knew for sure that I had a sasquatch standing outside my tent.

I replied that I didn't want to harm the sasquatch; I just didn't trust drunk humans because they were unstable and dangerous, and I wanted to protect myself from them.

I then had a vision in my head of my husband drunk in camp and berating me. I know it was a vision, and not just a memory of my own, because I could see my husband sitting in his chair and I could see Cristy and me also. I was sitting across from my husband, and Cristy was sitting next to him in front of the fire.

I then realized that I knew exactly what night I was being shown and I remembered how harshly my husband had treated me and the very unkind things he had said to me. I then said, "Yeah, like that! I don't trust them not to hurt me".

I took my hand off the stick while I was talking to it so that it would relax a bit and know that I didn't mean to use it on the sasquatch. About that time I heard the footsteps reach the top of my campsite and take one step down the stairs that led into camp.

Right then, I heard the man on the road gasp. I knew it was a man from the sound of his voice. He cried out, "What the hell is that! What the..." and he took off running down the road in the direction from which he had come.

I heard him reach his campsite and start yelling for everyone to wake up! He started pounding on cars while he was yelling. I could distinctly hear the sound of metal being pounded on.

I then heard a woman yelling back at him and saying, "Why the hell are you pounding on my car and waking me up?" She was really letting him have it.

He was yelling that there was something big on this mountain and they all had to leave! Now!

She told him that she wasn't leaving and if he ever hit her car again she would beat him up. I then heard that guy yelling, "Fine, you stay, but I'm out of here!" I heard a loud door slam and a car start up. He gunned the gas and tore down the road at a high rate of speed.

Where my campsite was located, there is a bend in the road and I remember praying that he didn't lose it on the corner and end up crashing into my campsite and running over me in my tent. When I heard him barrel past my campsite, I started worrying that he would crash on his way down the mountainside. It's a pretty rough road down this mountain with sharp corners and wash outs and, at the speed he was driving, any number of things could have happened to him.

In all this time the sasquatch hadn't moved a muscle and hadn't made a sound. When we could no longer hear this man's car driving down the road, the sasquatch took three steps back out of camp and went back into the forest.

I just lay there for the longest time trying to figure out what had just happened. I couldn't believe the Sasquatch would actually come to protect me. Did it know that the man meant me harm, or was it just a coincidence that it had been there at the same time that this man wanted to enter my camp? Or, had it seen the man leave his campsite and start to walk towards mine and simply wanted to know what he was up to? How would it have known that I was talking about my husband when I mentioned mean drunks?

I finally just accepted the fact that somehow it had known all of these things, and chalked it up to another amazing aspect of this species. All I did know was that I was very grateful for the help and very grateful for his presence at that time.

The entire camp left early the next morning and I was really glad that they did. For some reason, I never had any other contact from the sasquatch during my campout. I found that odd, but with them you just never know when they will come around, and when they won't.

Since I just couldn't stay away from this area, I headed back the next weekend by myself. By this time it was getting pretty late in the season and I knew my time up here was going to come to an end as soon as the snow would start falling in the next month or so. Already the nights were bitterly cold, and the sun barely warmed the air during the day. The sasquatch were a no show that weekend. I figured that had to be why I had met one on the coast in a semi populated area. They traveled down for warmer weather, and for the abundant food sources. I figured I finally knew which places this clan traveled to, and what time of year I would find them there. The more I learned about this species of life, the more respect I was feeling for them. They sure didn't act like the dumb animals that some people took them for!

Chapter 11 - I Must Say Goodbye

Personal issues forced me to relocate to Oregon at the beginning of 2010. I really didn't want to leave this area and I felt a great sorrow over it. I was having so much fun spending time in the mountains with this clan and I knew I was going to miss it. I never knew what crazy thing I was going to experience while I was up there. It really was a wild ride up to this point. It seemed that each time I was having trouble wrapping my head around something that I had read in a report, or just doubting it totally, I would see it happen here. I would then have to believe in it without a single shred of doubt! This place was my classroom and I really learned a lot from being there.

Take the orbs for instance. The first time I started seeing pictures with orbs in them and heard the people who took them state these orbs were real and not a blemish on the film or the lens, I doubted them... completely. I admit that... until I started getting orbs in my pictures while I was at this location. The first picture was perfect, the second had orbs. My camera hadn't changed and my lens hadn't gotten dirty. I had to admit to myself that yes it was possible to catch the orbs on digital cameras, but how can this be anything more than say…a burst of energy if you can't see it with your own two eyes? So I still had a shred of doubt left inside of me, right up until the time I saw the first orb with my own two eyes. The one that flew out over the top of the lake instantly vanquished all doubt in my mind as to the validity of orbs being real. As I've said time and again, you can no longer doubt anything you've already seen with your own two eyes. Especially if the occurrence lasted over an extended period of time and you had the opportunity to sit there and analyze it fully.

The same thing happened with the invisibility that many people associated with the sasquatch. Reports stated that the sasquatch can render itself invisible, or cloak himself at will whenever it felt threatened or just doesn't want to be seen. I read that they will use energy to create a shield, or something, around themselves that will shimmer like the Predator in that movie... and you have to look for this in the woods. When I first read this I thought it was the most far-fetched thing I had ever heard. I'll be honest here, I couldn't even begin to wrap my head around this behavior and wondered how anyone else could be so gullible. I mean, come on, this was something right out of Star Trek for goodness sake!

Then I saw the Orb flying across the lake simply disappear in front of my eyes. I sat on that log long enough to verify that it never splashed into the lake and it never hit the trees along the shoreline nor landed in the gravel. It never landed on the wooden dock that was right in front of it when it disappeared. I decided that I would keep an open mind but, truthfully, I still had that doubt and skepticism inside of me about this subject and I didn't truly believe that anything could just render itself invisible.

Since these events took place, I have also had other experiences that solidified my belief in invisibility. A lady, whom I trust fully, saw this as they were sitting beside me on the edge of a narrow river... but that will come later in the story. So, once again, I had to erase all doubt and I had to believe in something that I would have never thought possible. I hope that anyone who also thinks this stuff is simply impossible will just put a little more research into the subject of invisibility. I hope that someday they have the experiences they need to understand it also.

Before I moved away, I went up the mountain and camped alone one last time. I wanted to stop in and say goodbye to them, and just enjoy them being around my camp. I didn't hear any sounds at night so I figured they must have already relocated for the winter. I didn't see any recent sign of them in the area either, which really disappointed me.

Later that spring, I talked a new male friend of mine into going to this area to camp with me. I don't think this man had ever thought of the sasquatch before meeting me, and I really don't think he was a believer at all. He was a fellow lover of the woods and camping, so talking him into taking me up there was quite easily done. When we arrived, he agreed that this was a very beautiful area, and had a really peaceful feeling to it. The sasquatch didn't do anything to let us know that they were around camp at night, and the only really strange thing that happened occurred on the second day of our stay. At the time it really meant nothing to me, but this smell does come into play later on in my story.

It had begun to rain quite heavily that morning and never really did let up. The clouds hung low in the sky, creating a mist on the ground and it really looked like something from a movie. We decided to just hike around the lake instead of going up the mountain that day. There is a small foot path that follows the contours of the lake and, while we were walking along, we hit a narrow patch and I had to lag behind a bit so we could go through single file. He stopped suddenly and then turned to look at me and said, "can you smell that?"

I hadn't smelled anything out of the ordinary and was about to tell him so when it suddenly hit me. It was the unmistakable, and very strong, odor of dog poo. I mean to say it was strong! It was so strong that we actually checked our shoes to be sure we hadn't stepped in the stuff. Neither of us had any on our shoes, and I found myself looking around to see if I could spot a pile lying nearby. I mean, if the smell was any indication this would have been a very large pile of scat. We saw nothing around us and decided to just continue on down the path. As soon as we turned a few corners on the trail, the smell was gone.

As evening fell, it found us gathered under a tarp beside the campfire in the pouring rain. For some reason that I couldn't understand, this man kept putting more and more wood on the fire until it was bordering on a pretty good imitation of a bonfire. We had the tarp hanging over us in a "lean-to" style and I noticed the edge of it was actually melting at this point. I snapped at him and told him to stop adding so much wood to the fire. I told him that we were in a forest here, and I kind of liked the place just like it was. So he quit adding more wood but he never did tell me why he felt we needed such a big fire in the first place.

We retired early that night because of the rain and we were nestled all cozy in our sleeping bags just listening to the rain hit the tent. I was busy wondering where the sasquatch were and why they weren't coming around us that weekend. Suddenly there was a tremendous crash! It sounded like it came from right outside camp. We both jumped up in our bags and looked at each other.

He said, "What the heck was that?"

I had no idea. I had never heard something crash so hard in my life. Obviously, it had to be a tree, but that wasn't good. We were lying in a small tent in the middle of a forest with widow makers all around us. Not a good place to be, even if I was looking for the sasquatch. That was a really restless night for me.

The next morning we took a walk around the outskirts of camp looking for any trees that may have come down. We also wanted to make sure the roads were still clear to drive out because this sounded like a massive tree. We never did see any trees that had come down anywhere. We never noticed any large limbs lying around either. What made it strange was that the crash had sounded like it was so close to us at the time. We just walked back to camp and packed up since we were leaving that day.

I was so sad when we left because I thought it had been such an uneventful trip. After bringing him up here and us not finding any proof of their existence or having anything untoward happen in camp, I was pretty sure he thought I was just a nut job. There wasn't a thing I could do about this except to bring him back up with me again and hope I could find the proof he needed so he could quit looking at me sideways every time I brought up the subject of sasquatch.

A month later he and I met Cristy up there, and we had such a wonderful time. The weather was perfect and the company was great. The sasquatch, however, were a no show again. We did make a great discovery that would come into play later. We walked up the logging road out of camp and headed to the top of the mountain. When we crested the top we came into a heavy forest and was in shadows most of the day. The scenery was so lovely and I kept taking pictures

Figure 21 White Orbs

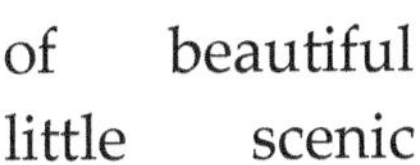

of beautiful little scenic

spots we found along the way. The camera wasn't wet or dirty but I kept catching assorted sized white orbs, blue orbs and even a very large orange orb in various spots through here. None of these were seen with the naked eye.

The game trail that we were following dumped us into a long, rectangular field. After being in the shade of the forest for so long, I had to let my eyes adjust a bit before I could see where I was. The place was simply beautiful. The sun was warming the waist high grasses and at the very end of this large field was a great fallen tree which was sun bleached to a beautiful silver color. As we were walking through this field, we were keeping our eyes and ears open for the sounds of anything that may have been enjoying the sunshine here. The grass was so high there were easily hundreds of places where something could have been lying in wait for us to walk up to it.

About half way across this field, Cristy called me over to where she was standing and looking down at something on the ground. As I approached her she told me that she thought it might be a bed or something. I looked down and sure enough, I could see where something very large had been laying. We could see the imprints of its head, waist,

Figure 22 Friend on large tree in field

and even the imprints of where its legs had been extended. We were both so thrilled to see it and started talking about how, if we lived on this mountain, this is exactly where we would come to get some sun. We walked across the field and over to the big tree to check it out.

It was really impressive and we all had to crawl up onto it.

At one point, I wandered away from the tree and started to look around in the bushes and shrubs that grew around the perimeter of the field. I spotted a big pile of large bones and went over to see if I could identify them. The more I walked through this area, the more bone piles I found. Some of the older bones piles had obviously been scattered by other animals, but it was the piles that captured my attention the most. I picked up a very large weathered bone that looked like it could have been part of a spine. I took it home with me because of its unique shape. I remember wondering if this was a central meeting place for hunters to bring their elk kills, or if this was a place that other animals brought their kills. I never did figure this one out but I decided that I would name this field. I walked back over to the dead tree and proclaimed this field to be the "Bone Field".

At one point that night I walked out of camp and into the darkness of the forest to relieve myself. While I was walking back towards camp I noticed that I could see everything that was going on in camp very clearly since the campfire was lighting the area up so well. I stopped where I was and thought about how easily it would be for a sasquatch to just stand outside the fire light and watch us. I wanted to do an experiment and see just how close to camp I could get before anyone noticed that I was standing there. I was wearing the same clothing that I had on when I walked away, and none of it was black, so I figured that someone was going to spot me well before I reached the outskirts of camp.

I walked up to the first tree trunk and hid behind it, and watched camp for a few minutes. No one noticed anything so I walked to another one and did the same thing. I walked from tree to tree always stopping behind them and peeking out at camp before continuing on my way. I figured this was the way the sasquatch would do it. I could hear them talking in camp quite clearly, and I would stand and listen to a few sentences before continuing on to the next tree closer to camp. I had now arrived at the tree that sits right outside of camp and hid behind that one. I had been crunching the branches and needles on the ground the entire time I was doing this, since I hadn't learned the art of walking silently through the woods yet and I thought for sure that one of them would have heard me stalking up on them.

Figure 23 Our secluded campsite

I stood there looking around the side of this tree wondering just how they had missed all the sound effects and hadn't noticed me standing there, or even felt me staring at them so hard. I was wondering if perhaps they were just ignoring me when I heard Cristy ask, "Where is Kathi, shouldn't she be back by now?" I stood there for a few more minutes and then just walked into camp normally. I didn't tell anyone about the experiment I had just conducted. I didn't want them to know what I did just in case I might want to conduct this experiment again in the future.

I had a newly-found respect for the sasquatch then. I also thought I knew exactly what they were doing up there. It was so easy to sneak up on my friends sitting there. Staring into the campfire totally negates our ability to see in the dark which leaves us very vulnerable in the woods at night. Now, when I am camping alone, I try to keep my back to the fire in order to keep my night vision sharp. I don't want anything sneaking up on me without my knowledge.

I was mildly concerned about the fact that I had now come back to this area twice and I hadn't had any interactions with them, nor had I any sign of them being in the area. I pondered this silently on the drive home, and I was really hoping that nothing bad had happened up here that forced them to leave this area.

Chapter 12 – Memaloose Trail

When my husband and I separated, I moved in with Cristy and she became my research partner. After the things we had seen and done together she was becoming as obsessed with the sasquatch as I was. We shared a beautiful apartment in Clackamas, Oregon, and I was finally able to research and experience the sasquatch without constant criticism and ridicule. I could never understand how my husband could have had his sighting out in the marsh, and then completely lied to all of his friends about it. I had a sighting and became obsessed. I wanted to know everything there was to know about this marvelous species of life, and I had no problem sharing it with everyone around me.

I began to tune into blogtalk radio every single night to listen to any show that featured stories about sasquatch. I actually sat and listened to every single one of them just to get any new tidbit of information that I may not already have had on the subject. Oldie One was also hosting a radio show, and he approached me with an offer to co-host for him. I'm a very private person normally, and I really don't like to talk in front of people, but because he was a friend I gave in to him and co-hosted a few dozen radio shows.

I met some really great people during that time, and I learned some very helpful things about the sasquatch and their behavior, but I found that no one really had any new information to share with me about the Twilight Zone type of experiences that I was having. The few times that I attempted to bring them up, you would have thought that I had just said that I was an Alien from another Planet! I heard so many people say that I was just making this stuff up to get attention or that I must have been high and hallucinated all of this stuff, when nothing could be farther from the truth!

I told Oldie One that I could no longer be a part of his show, and to please find someone else to take over for me. I wasn't tough enough at that point to handle all the slander and ridicule that I was getting just because I came forth with what was happening to me out there in the woods. It didn't keep me from going out into the field anymore... it just kept me from ever sharing with the general public.

I just stopped reporting anything, and I mean anything, that I was finding. I continued to write in my journal each time I went out, but I quit trying to get pictures, and I quit trying to record anything. I decided that I wasn't in this to try to convince anyone. I was in this to enjoy every second that I was allowed to spend with the sasquatch. There were enough people out there recording sounds and trying to get pictures of this species. I would let them race to the finish and see who could prove the sasquatch existed and what they truly were. I wasn't in this to get rich or find fame. My experiences were something that I felt didn't need to be shared with anyone else, except Cristy and a few close friends.

Oldie One did introduce me to some really fine people while we were doing the show, and a few of them became really good friends to me. These were the only people that I was sharing my experiences with. I guess, to be honest here, I got angry. I knew that I was not lying about any of this, and I did not appreciate the fact that people who didn't even know me, and had never even met me, were treating me this way. And besides, I had even greater experiences than the ones who were calling me names had anyway. So I figured they were all just jealous, since most of them had never had an experience at all, and the rest had only found an old footprint on a trail somewhere. How could they even understand what I was going through or seeing?

One day in early spring a friend and I were riding on his bike and we discovered a road called Memaloose Trail. It looked like it would be a perfect area for the sasquatch to hang out. I wanted to go back and check it out. A few days later we drove back up in his van and proceeded up the mountain as far as we could go. It was around April and the roads at the higher elevations were still blocked by snow that hampered our progress. On our way up the mountain we passed an officer who we stopped to inquire about camping in the area. He told us that it was indeed legal to camp anywhere up here, and there was a lake up above with available primitive camp spots also.

We decided to go on up and check out this lake, but before we got there we found that the road was still covered in snow and ice. A jeep stopped us to say that the road was completely impassable just up around the next corner. We pulled over at that point and got out and walked around for a bit and I decided that I was definitely going to come back up here and camp later in the summer. It looked to be a very promising area.

I then found out that there was going to be a conference in our area called the Oregon Sasquatch Symposium, and it was going to be held in Eugene, Oregon. All the great's in the Bigfoot community were going to be there. Even Mr. Bob Gimlin himself was going to be in attendance! I had to be there, especially since I viewed all of these people as being the cream of the crop when it came to Bigfoot research. I had about a zillion questions that I wanted to ask each and every one of them, and Cristy and I quickly made plans to attend.

I was a bundle of excitement, and nervous energy on the drive over. I planned to just sit and listen to what everyone else had found, and not divulge too much information about what I was doing out in the field. When we finally arrived, I wanted to go straight into the bar and order a little something to calm my nerves before heading to the meet and greet, so Cristy and I grabbed an open table and sat down.

While we were sipping on our drinks, I glanced over at the next table and just about wet myself. Sitting there were two people that I had elevated to celebrity status in my mind. Mr. Bob Gimlin and Dr. Jeff Meldrum were sitting at the next table talking to each other! I grabbed my phone and texted my ex just to brag to him that I was in the company of a certain Mr. Bob Gimlin at that moment. He wasn't too impressed, but I was like a kid in a candy shop. Of course, I was much too shy to just walk over, or lean over, and introduce myself to them, and they caught me sitting there just staring at them like I was a stalker or something. I grabbed my drink and went out on the balcony where I could continue to stare at them without getting caught.

Of course we both had our pictures taken with Mr. Gimlin at the symposium. He is the Godfather of the bigfoot world in my eyes. He and Roger Patterson were the first ones to ever get a very clear video of the sasquatch walking away from them, and unfortunately people are still arguing over the validity of the film today, but I'm convinced it's the real thing.

The information that was shared during this event was fantastic, and we had a wonderful time, but it was all mostly mundane experiences and nothing from the Twilight Zone was ever mentioned. I learned something very unfortunate during this Symposium. I learned that most of the people who thought of themselves as "experts" in this field, and who actually looked down on Cristy and me and made us feel small, had not only never seen a sasquatch, but they had never even been close to one!

Figure 24 Cristy and I with Bob Gimlin

I found out later that the self-proclaimed "expert" who snubbed me the hardest at this venue had never even found a sasquatch foot print in the ground, much less had any real interactions with them! How the heck can you be an "expert" at something you know nothing about? And about something you have never even encountered? That really got my goat. But some great things did come from attending this Symposium, don't get me wrong. I met some fantastic people and we have stayed in touch ever since.

I was also able to watch a demonstration on the proper way

to cast footprints here, and I recognized everything that my husband and I had done wrong the first time we attempted to cast the hand print. I found out that to glue the sandy soil together and get a perfect cast, you must first spray the print lightly with hair spray. Also that you could build a box around a print that was left on a hillside and still get a perfect cast without the casting material flowing down the hill and away from the print. That was very helpful to me and I enjoyed it immensely. The Symposium ended much too soon, and I was pretty unimpressed with most of the people who I had been told to look up to in this field. I decided that I had made the correct decision in backing out of the Bigfoot community.

I don't want to leave the reader with the impression that I didn't like anyone that I met at the Symposium! A lot of the attendees have since become very good friends of mine. It was just the few who seemed to have a super inflated sense of themselves that got on my nerves. I assumed that in order to have such an inflated ego, one must also have the knowledge to go with it. I found this to not be the case eighty-nine percent of the time.

After we attended the Oregon sasquatch Symposium, Cristy and I decided it was time to go on up to Memaloose Trail. We wanted to take a hike up to the lake and do some camping, so

Figure 25 Hiking to the lake

we loaded up her truck and hit the road. It was a beautiful summer day, and the sky was a brilliant blue.

It was a perfect day for a hike up to the lake and for camping, but unfortunately, everyone else from the city had the same idea. Neither one of us likes to camp where others are situated close by, so we talked it over and decided to just drive farther up the mountain and find ourselves a nice quiet spur road somewhere to camp by ourselves.

We followed road after road searching for the right spot but we weren't going to give up until we found it. We finally stopped in front of a narrow road that curved off to the left so we couldn't see the end of it, but it looked promising. We found that this road ended in a big circle covered in gravel. The back side of this circle dropped down into a steep canyon with a perfect view of Mt. Hood to the left of us. We parked the truck and walked over to the edge to check it out. We instantly knew that this was the place we were staying. Right below the drop off were some huge boulders that jutted out over the canyon and would be just perfect to sit on and drink our morning coffee and

Figure 26 Boulders below our camp

also sit and enjoy the sunset.

After we got camp all situated, we headed down to the boulders. We noticed a pretty well used game trail that led from the top of our overhang down into the bottom of the canyon below us. We also noticed that there were a lot of things laying around us that pointed to a high rodent and chipmunk population. We threw out some grub for our little brothers to snack on later.

I had brought along my binoculars and was sitting on top of the boulder looking out across the area when I spotted an owl sitting at the top of an evergreen tree. It looked so pretty sitting up there and I decided to try to get a picture of it, so I lowered the binoculars and grabbed my camera. That was when I noticed how far away it really was, and my camera wouldn't zoom in clearly that far away. Cristy and I started experimenting at that point. She figured out that if she put the binoculars up to the lens finder on the camera, she could get some pictures of the owl that way, and it was actually working.

Then the owl took to the sky and began to fly directly towards us but it never got any bigger in size! It landed in a tree that was only a few yards away from us, and that was when we noticed that this was the tiniest little owl that we had ever seen! It was also the cutest little thing we had ever seen! It was about the size of a sparrow, yet it looked just like an adult owl. After we got home I looked this little thing up on the internet to see what its name was since I had never seen an owl this tiny before. It was a Pygmy Owl, and apparently they are quite common. I sure hope I get to see another one someday.

After sitting on the rocks enjoying the sunshine and the view, we decided it was time to head back and check out the rest of our little camp spot. On the other side of camp, across from the canyon, was a slight hill which we noticed was being used as a backdrop for target shooting. There were pounds of discarded shell casings, broken glass, cigarette butts, beer cans and other human paper garbage littered around this small area. We gathered all that we could, and placed the burnable stuff into the unused fire pit. Seeing so much litter lying around here made both of us so mad we were seeing red! Of course we did our share of grumbling about it as we lit our campfire that evening! We were in unanimous agreement that humans are definitely ruining the earth, but at least for now our little campsite looked a whole lot better.

When the sun set that night it was extremely cold on the mountain and we huddled in front of the fire trying to stay warm until well after dark. We always bring along a portable CD player whenever we camp so that we can enjoy listening to our favorite Native American musicians in the forest where they were meant to be heard. We absolutely love the sounds of a cedar flute and drums wafting through the trees.

Nothing of note happened that night and we went into the tent and got into our sleeping bags, but it was so darn cold that neither one of us could sleep. We tried putting some rocks from the fire pit into a pan and bringing them into the tent with us to keep warm, but what little heat they were giving off was being lost out of the tent as fast as we were feeling it. We ended up pulling out all the extra clothes we had brought with us and we laid them out under our sleeping bags for an extra layer of warmth between us and the ground. I was ever so grateful when I finally saw the sun come up over the mountain.

The next day we decided to do some sightseeing while we were in the area. We drove up the main road and then continued to check out every road and every spur road that exited off it. We drove down one spur road and came upon an area that was so beautiful we decided to get out and hike through the forest there so we donned our packs and set out. While we were walking through the woods here, we stumbled onto a human created lean to, which we sat in to take a well-deserved rest. It was very well made, and it blended into the surrounding woods so well that we almost walked right by it.

After walking around this area, we jumped back into the truck to head farther up the mountain. As Cristy was driving along as I was looking off to the side of the road to see if I could spot a nice area I might want to check out. I noticed a clearing at one point and asked Cristy to pull over so we could check it out. When we walked down into this clearing, I couldn't believe how beautiful

Figure 27 Orbs in clearing

it was. The ground was covered in Moss, and it was bordered on all sides by the trees. Cristy remarked about how peaceful it was in here and we walked around for quite a while in this area. As we were walking out, I stopped to take a picture to remember it by, and later found all kinds of Orbs in the picture

When we got back to camp we walked down to the boulders to enjoy the last of the sun and then catch the sunset. I sat on the upper boulder and Cristy sat down below me on another boulder when she pulled out her flute and started playing. It sounded so peaceful and soulful echoing back to us from the canyon walls. I just sat back and listened to her play while watching the sunset lighting up Mt. Hood.

When we decided to go back to camp, we laid one sea gull feather and two white rocks on one of the boulders where we were sitting. There was one white rock from each of us as a thank you. We scattered some more food for anything that was hungry and climbed back up the hill. We started a fire and ate our dinner huddled up in front of it trying to keep warm and talked about the sights we had seen that day. I pulled out the new harmonica I had brought but couldn't seem to get any real music to come out of it so I gave up. At one point in the evening I remember Cristy saying that she was going to go sit in the truck, and I decided to join her there to ease my aching back.

It was fully dark out now, and there was no moon so the only light in camp was a small circle around the campfire. Cristy was laying back in the driver's seat, and I was sitting up in the passenger seat with my side window facing the campsite. As we sat there talking, I don't remember if I closed my eyes or not, but I suddenly had a vision go through my mind of something looking in the side window.

I instantly whipped my head around to look out the window and there was nothing standing there but as I was looking into camp in the direction of my tent and the trail that led down into the canyon, I thought I caught a slight movement from in the dark. I felt very lightheaded suddenly, like I was going to pass out or something. I sat back, turned my head forward and started looking out the windshield. I think I closed my eyes for a second. I then started feeling really scared for no reason at all, so I started talking to myself and trying to calm myself down. I was thinking that I had just made myself nauseous by whipping my head around so fast like that and I had no idea why I was trying to scare myself. I'm not exactly scared of the dark, and I didn't want to feed the feeling to Cristy, so I didn't say anything to her about it at the time.

I noticed a shadow fall on the side window next to me, but I was unable to turn my head to look in that direction at that point and wondered why. I remember wondering if I was just trying to scare myself or was there really something out there. I noticed that my midsection and legs were feeling very heavy, like a weight had been placed on them. I was finding it hard to breathe then and I decided that I just needed to get out of the truck, and get some fresh air, so my stomach wouldn't feel so queasy.

I wanted to reach for the door handle and open the door, but I found I couldn't lift my right arm into the air. I tried to lift my leg off the floorboard of the truck at that point, and found that I couldn't move it either. I don't know exactly how long this episode lasted, and I never mentioned a thing to Cristy about what was happening to me until I started babbling. I remember saying, "We have to leave right now! No, I have to leave right now! I don't want to leave right now! No, we have to leave right now!"

At this point the episode ended and I felt this weight lift off my body, and then my legs felt like they were floating. While I was analyzing this new feeling of weightlessness, Cristy said in a real small voice, "My legs feel so light!"

I couldn't believe she had just said this and I exclaimed, "My legs are floating!"

To which she replied, "No, they feel light!" I told her that wasn't the point. I called it floating and she called it light, but what the heck had just happened? She told me she could hear me talking, but couldn't reply, or even move, at that moment in time to find out what I was talking about. She was feeling the same things that I was feeling! We couldn't have fed each other the information because we never talked about what was happening to us until after the experience was over.

We sat and discussed what we had just gone through, and tried to figure out exactly what had just happened to us, until we noticed that the fire was dying down and we sure didn't want that to happen. We decided to get out and while one of us was putting wood on the fire, the other would get the sleeping bags out of the tent, and we would just bed down in the truck for the night. This sounded a whole lot safer than sleeping in the tent at that point.

Before we got out of the truck Cristy hit the alarm button on her trucks remote. When she did this the horn honked repeatedly, and both the headlights and taillights flashed on, illuminating the front and back of the truck and pretty much the entire camp. We didn't see anything hanging around the truck, or in camp, so we figured we'd make a run for it and do what we needed to do.

Figure 28 Our Camp

When we finally talked ourselves into getting out of the truck, the campsite felt peaceful and calm and so did we so we just stood in front of the fire warming ourselves and discussing what had just taken place. Neither one of us could really put a finger on it, but we were feeling better all the time and decided to just go ahead and sleep in the tent. I don't really remember waking up again that night and I don't think the cold bothered me as much as it had the night before. We took down camp the next morning and drove home.

I kept thinking that if we would just keep going back up this mountain and camping, and playing our music, they would come back in and not be so mean and we could form a trusting relationship with each other. (If that was what was up on the mountain with us) This is exactly what I did with the Washington clan, but I had never felt this type of terror up there with them, and I camped up there alone. I'm not sure I would ever want to camp up above the Clackamas River alone.

Yes, I now thought of them as a clan of semi humans, and not a family of apes. The experiences I had with this species changed my mind. I knew they traveled in family groups because I had seen the trackways. I knew they had compassion because they had protected me. I knew they ate with their hands because of the paper plate incident, and I knew they were quite grateful for the things I had given them, because of the feather. None of that shows me an animal side, it shows me that they articulate thought and are capable of love.

For those that are all wrapped up in the thought that man is the dominant species because we have a thumb, well…they have a thumb! The only real difference between us and them is the way they live, and the amount of hair on their bodies. They have never adapted life to suit them, they live in Nature and they do it quite well. Perhaps they never wanted to live as we do in our climate controlled dwellings, with petroleum product fumes assaulting us from our paint, our carpets, linoleum, clothing, shoes, etc., and the sound of electricity buzzing through our walls. I don't think they ever will, and I can't blame them for that. I have always had the desire to walk off into the woods and never see a city again, so perhaps I just understand how they might feel a little better than the average city dweller.

We did have a chance to return to this spot, but we didn't do it alone.

Chapter 13 – Return to Memaloose Trail

I met Thom Cantrall when Cristy and I went to the Oregon Sasquatch Symposium, and I felt that he could be trusted. He's a great big cuddly teddy bear with a huge heart, and his intelligence and wit will keep you on your toes! He's like the big brother I always wished I had. I called him to discuss the strange things that had happened to us up on Memaloose Trail and it was decided that he would drive down and camp with us there and check it out. If nothing new happened, at least we would get a camping trip in together.

Figure 29 Thom Cantrall and I at OSS

We all met on the side of the road on the way into this area and we laid out a map over the hood of Thom's car to try and figure out which would be the quickest and easiest way into this little camp spot. It turned out that we could make better time if we went in the back way, instead of the way Cristy and I had gone in the first time, and that's just what we did.

The drive in wasn't as nice as we had assumed it would be, however. The further over the mountain we drove, the more our eyes were assaulted by human garbage lying on the sides of the road. There were beer cans every so many feet, recliners, dryers, bags of household trash, soft drink cans, and soft drink cups, McDonald's bags, and Burger King bags. Enough to make you just want to burst out in tears! It was hardly the beautifully scenic drive through the countryside that we were all expecting.

When we finally arrived at our secluded, out of the way spur road I was very happy to see that no one had been here to dump more garbage around this area while we were gone. That was a relief because I really didn't feel like having to clean that spot up again!

We got busy raising the tents and getting camp all set up just the way we wanted it. It was a beautiful, sunny day and we all had a nice walk around the area while Thom gave us a quick lesson on the names of each species of tree that was growing there. He's a great teacher, and very patient.

We spent most of the day driving around some of the roads on the mountain taking in the scenery while also gathering firewood. We were all taken aback by the amount of garbage left in this place! We couldn't pick up all the batteries, and other toxic garbage, that was left lying around everywhere, but we collected as much as we could. I will never understand why someone would come out into the pristine beauty of nature just to dump their trash. I just get so angry it makes my stomach churn. We did find some incredibly beautiful places on our journey, which made up for some of the ugly garbage we found.

Figure 30 View from Memaloose Trail

When we got back to camp it was time for dinner and relaxation around the campfire. We shared a very nice meal of Thom's Famous Chicken Hips, and then he serenaded us around the fire. Thom is quite the crooner and has a fantastic singing voice and I could have listened to him all night. We kept him singing as long as he would participate, but it ended all too soon. The poor man deserved his rest as we had all had a very busy day collecting firewood and garbage.

When the weather began to turn too crisp and cold for us to be sitting out in it any longer, Cristy and I retired to my tent and Thom climbed into his. Cristy fell asleep soon afterwards, but I just wasn't tired enough yet. The back window of the tent was again facing the drop off, so I unzipped it and sat on my sleeping bag looking out and listening to the sounds of the night. I really enjoy doing this at the end of the day, just listening to the quiet and reflecting on the day's events.

As I sat just gazing out the window and listening to Cristy snoozing contentedly, I heard the sounds of something coming up the side of the hill. I couldn't see anything and the only sounds I heard were the sounds of the vegetation rustling in certain spots. I was hoping to spot a coyote or maybe even a cat walking out of the ravine. To my surprise it wasn't anything that small that topped that hill and looked into camp.

It stopped by my tent and I felt a sudden rush of fear overtake me. I said out loud, "Oh no! You're not going to scare me this time! I'm not here to harm you and I'd appreciate it if you don't harm me either!"

I heard a conversation start up in my head and physically shook my head to lose it or perhaps to make sure I wasn't making it up myself. The voice continued and it sounded really angry. It said, "You have no respect for your home! Go away!" I replied that I had a great deal of respect for my home! Then I instantly thought that maybe I did know what he was talking about and thought about all the litter we had seen that day. It was everywhere and I was just as appalled at the way we humans were dumping our garbage in the forest as they were. The sasquatch live here, so we're actually littering their homes and they don't seem to be very happy about it. I couldn't deny a thing that he said!

As I sat there listening to this, I just wanted to cry. I knew that they were suffering because of us and I knew that I never wanted them, or their children, to suffer. Humans can be so darn careless and closed hearted to the wants and needs of other species, and here was the best proof of that I would ever get. I just wanted him to know that we weren't all like that. That some of us humans do put the needs of other species in front of our own, and that we do care about Mother Earth and all who live on her.

So I began to talk to him. Not in my mind, but out loud. I figured that if I did wake Cristy up then she could be a part of this. I replied, "But we here in this camp do care! We have also seen the garbage and filth that is ruining this beautiful mountain and the swimmers homes. Just look into the back of our truck, the thing we came here in. We've been picking up trash all day and we plan to take it out of here with us. We couldn't get it all because there is so much of it, and we've run out of room. I'm so incredibly sorry that most humans don't know better. Please know that we are not like them! We want Good Will, and we would never do anything to harm you or your children!"

By this time I felt so bad for them. I would be angry too, if I had to see this sort of behavior happening over and over again and could do nothing about it. I couldn't see him directly as he was standing on the side of the tent and I don't have windows on that side, but I heard him start walking across the gravel towards Thom's tent, so I figured he was done with me and was going to go mess with Thom for a bit. Since I was absolutely freezing by this time I decided to lay in my sleeping bag and listen for any other sounds from that side of camp. I must have been very tired by that time because instead of staying awake and listening, I lay my head down and fell fast asleep.

I know everyone is asking themselves why I didn't get out of the tent, or even stay awake, in case Thom was threatened by this creature. I never once felt that Thom would be threatened. I knew that Thom could handle him better than I could and I was never threatened in any way from this creature. It was angry, but it wasn't about to start ripping our heads off over it. It was concerned, angry and wanted to be heard. We heard it and we replied to it. There was really nothing else to be done, and I sure didn't want my actions to be taken as defensive, or offensive, in any way.

I woke up refreshed and noticed that I had barely moved in my sleeping bag. I can tell because when I toss and turn in my bag, the zipper ends up somewhere besides where it was when I zipped it up the night before... usually lying across my face. When I fell out of my tent both Cristy and Thom were already awake. I didn't say anything for a while because I wanted to see if anyone else had anything to say about our visitor. I finally found myself just blurting out that we had a visitor come into camp the night before, and had anyone else noticed. Thom just smiled and said, "I know" and I told them what happened at my tent, and that it had walked out towards his tent when it left mine. I hadn't known that Thom had spent the night in his car until he told me right then.

Thom replied that he had been sleeping in his car and woke up to the feeling of pressure on his chest, pressing him down as if a weight had been put on his chest. He said that he was struggling to breathe and couldn't raise his arms. The same thing that had happened to us when we were sitting in the truck! Thom stated that the sasquatch had told him how concerned he was with the fact that humans kept leaving so much garbage lying around, and how we really have no respect for our homes at all. Then it told him that they didn't live in this area, they were only here to catch the swimmers in the waters. Thom says he also told the sasquatch to look into the truck and see the amount of garbage that we had picked up and were taking out of this place, and the sasquatch left him with the question, "Do humans not know respect?"

So we had both had the same conversation with this fellow and heard the same things, and neither one of us knew about the other's experience while it was happening. We talked about this for a while and I felt good about the fact that Cristy and I always cleaned up the garbage from every area we ever went into. Seems like such a small thing to do for the Earth, and I believe the sasquatch greatly appreciate it.

Thom said that it really was too cold for him to stay there another night so we helped pack up his gear and he left after dinner that evening. Cristy and I stayed on another night, but nothing else happened to us, and nothing entered camp.

A few weeks after our trip up to Memaloose with Thom, I started having some really strange things happen in my bedroom and then out in the rest of the apartment. It all started one evening while I was lying on my bed listening to music and I felt someone's hands on my head. It felt just like someone was standing behind me and they were putting one hand on each side of my head and the fingers were pressing in on my temples. I distinctly felt one hand on each side of my head. At first it was just a light pressure on my temples. I just lay there feeling it and wondering if I was about to get a headache or something. Then it began to push harder and harder on my temples, until the pressure started getting to be too much and it felt like it was going to pop my head like a bubble or something. I jumped up off the bed and looked down... of course I couldn't see anything and the pressure had released as soon as I jumped up. I stood there next to the bed for about three minutes trying to figure out what had happened and eventually decided that it must have just been pressure in my head from lying down and I pretty much forgot about it soon afterwards.

We then began finding our front door unlocked when we came home. The door would still be closed but both the doorknob and the deadbolt would be unlocked. The first time we noticed it, Cristy had come home first and found it unlocked. She blamed me for leaving it unlocked all day. I assured her that I never left the front door unlocked, and we argued about it a little. The second time, I came home and thought that she had left the door unlocked and we argued a little bit more. This kept happening until the final time when Cristy tried to blame me for leaving the door unlocked and I knew for a fact that it was locked behind me when I left. The reason I knew that I had locked it behind me was because my friend had locked the door behind him when we walked out, and I had leaned over to double check it to be sure, so we both knew beyond a shadow of a doubt that both locks on that door were locked when we left. We knew then that it wasn't either one of us that were leaving the door unlocked.

Nothing was ever missing from the apartment when the door was left unlocked, so we couldn't figure out why someone would come along and unlock the door but not take anything or seem to move anything. This happened so often that I even went so far as to ask the management if someone else could possibly still have a key to our door or if maintenance was coming into the apartment during the day. They said they didn't think anyone else had a key and assured me that maintenance wasn't coming in so I asked if they would just replace the lock for us to make sure. They did come out and change the locks and after that I had no explanation for why this kept occurring.

On another night, I was sitting on my bed listening to one of the Blogtalk radio shows on my laptop and typing in the chat room. I had a double closet in this room with mirrored doors that lined one entire wall of the room and my bed sat across from them. As I was sitting there I felt a hand rubbing the back of my head, starting at the nape of my neck and then moving upwards towards the top of my head. I instantly looked into the mirror for whatever reason but of course there was no one standing behind me. There couldn't have been anyone behind me because my bed was right up against the window and I was on the third floor. I didn't see anything and the feeling stopped as soon as I had whipped my head up from my laptop and looked into the mirror.

I figured it must have just been my hair sliding across my head and continued typing into my laptop. Then I felt the hand rubbing the side of my head and a big wad of my hair was raised up into the air. I again looked into the mirror and this time I could see my hair hanging sideways in the air and no one was on the other side of it holding it up! This scared the heck out of me and I instantly grabbed that bunch of hair and pulled it down again. I then began waving my arms around my head and through the air striking out at whatever the heck was in my room with me, and saying, "Don't touch me!" I ran into the living room and sat on the couch in total disbelief of what had just happened to me. Cristy wasn't home at the time so I called a friend and actually stayed at their house for the night because I didn't want to be in our apartment alone.

I calmed down after a few days and went back home to sleep again. Everyone was telling me that ghosts don't exist and even if they did exist they couldn't hurt you. I didn't know about that one, but it was time to get on home anyway. I tried not to think about any negative thoughts when I got there. I made a pot of tea and listened to a little music to calm myself before getting ready for bed.

Once again, I felt the hands on either side of my head when I laid down. I laid there trying to analyze what was going on and trying to feel in my brain to see if I was being scanned or something. I really had no idea why only my temples felt the pressure getting tighter and tighter even though I could feel the weight of the hands, or whatever, on both sides of my head also. Only my temples were being squeezed. I took it for as long as I felt it was a safe pressure. Once it got too tight for me, I once again jumped up off the bed. This time I spoke out loud and said that whoever this was they were not given permission to be in my room, or to be assaulting my body in any way, shape or form.

I never felt a presence when this was happening, or saw anything at all, but I knew that something had to be there in that room with me. I decided that I was not staying here by myself, and once again sought refuge at a friend's house. While I was there I got a call from Cristy asking me to please come home right now. She was very adamant that it was not to be tomorrow morning; I was to come home right now!

I'm going to stop here to introduce the layout of our apartment to give a clearer understanding of what happened next. We had a long hallway that came out from the living room and ended at my bedroom door, with Cristy's door off to the left and my bathroom door off to the right. You could sit on the couch in the living room and look down the hall and see my bedroom door at the end of it.

When I got home Cristy told me that she had come home and slipped into a warm bath. While she was sitting there reading in the bathtub, she heard the front door open and close and someone walk down the hall past her bedroom and into my bedroom. She thought that I had come home and she called out to me, but I never answered her. When she got out of the bath she opened my bedroom door so she could talk to me but I wasn't in there. Then she looked around the apartment for me and was very surprised to find that I wasn't even home. She sat down on the couch and was wondering if she had really heard someone in the apartment while she was in the bathroom. Then she saw my bedroom door open and close! That was when she called me and ordered me to come home.

When I got home and heard her story, I told her about the things that were happening to me. I had told her a bit about what I was going through when it was happening, but I don't think she really understood how unnerving these things were until they started happening to her also. Nothing else happened that night and the only thing we could think of was perhaps we had a ghost in the apartment. I asked the management if anyone had ever died in our apartment in the past. They said no and wanted to know why I was asking. I told them I was just curious, but I'm sure they thought I was totally strange for asking. This was the only thing we could think of to explain what was going on during this time so we got some sage and we smudged the apartment from one corner to the next to try and get rid of the ghost.

Everything was quiet until a few days later when a friend came over to visit. Cristy and I were sitting out on the patio enjoying the sunshine and I could see through the windows into the living room. I watched him walking through the room on his way out back and then he suddenly stepped off to one side and turned around really quick. I went up to him and asked him what that was all about, and he said that something had touched him on the back and he had turned around real quick to see who it was. Of course there was nothing behind him, and he wanted to know what was going on. I reminded him how I thought the house was haunted and all the strange things that kept happening to me. He said that until he was touched he really hadn't believed me.

About a week or so after we swept with the sage, I was once again sitting alone in the apartment. I was in my room with the door closed when it suddenly filled with the overwhelming scent of poo! The first thought that filled my head was that one of Cristy's cats must have come in and defecated in my bedroom. They never did do that or go anywhere other than in their litter box, but it was the first thing I thought of at the time. I got up and circled around the entire room looking for it. I looked under the bed, and then through the walk-in closet. Surprisingly the closet still smelled like fresh clothes so I came out and closed the doors behind me. I looked under every table I had in the room, and I ended up at my dresser. I looked under the dresser and then stood up next to it. As I stood up the smell suddenly changed from a strong poo smell to the scent of oranges and blossoms. I was stunned, let me tell you! Now my room was filled with this citrus and flowers smell, and it was a really nice scent!

At that point I just knew that it couldn't be a ghost we were dealing with and I wondered if the sasquatch could follow you home and appear in your room. I remembered when my mother passed and I first found the sasquatch. For three days I would be crying in my room and I would feel a presence appear. The energy in the room would change and something would take my hand, and I felt so comforted. It was a large hand that would encompass mine and a wave of love and compassion would just pour over me. I had never felt so loved up to that point, and I've never felt so much love since. I could never understand what was happening to me while this was going on but the feeling of being comforted felt so good and I felt so calm while it was happening that I just enjoyed it at the time. I never knew if it could be my mom or another entity coming in.

So, I had a feeling that this had to be the sasquatch in my room now, and with me when I cried back then. I felt super bad about smudging and trying to get rid of them, and I hoped that they would come again so I could show them the respect they deserved. Unfortunately, I never had anything touch me again after that, and nothing strange happened in the apartment.

I think the smells were a way of letting me know that it had been them in the apartment and not a ghost at all. I also think that it was a goodbye message of some sort since we never had anything strange happen after that.

Chapter 14 – Summer Adventure

I moved to back into Washington state in the beginning of 2011. Cristy had become pregnant and wanted to move the baby's father into our Clackamas apartment so it was time for me to move on and let them start their family. I ended up with a real humdinger of a roommate this time. I won't go into too much detail about it, but my home life was really bad while I lived there. I never had another strange encounter in my bedroom, but I did start picking up orbs when I took pictures in this house.

Figure 31 Orbs in the living room

I attended my second Oregon Sasquatch Symposium that summer. It was being held outside of Sisters, Oregon that year and I drove down with Thom and his friend, Charles. It was a long drive there from Washington but it was shared with great company and we all enjoyed the drive very much. In order to avoid the large crowds that were staying at the venue, we chose to stay in a separate campground just off the property rather than staying in the cabins with everyone else. We picked a nice camp spot situated next to the river that had a nice hiking path leading off into the trees.

After we got our tents up and the camp arranged a bit, the two men went off to the venue for the meet and greet. I stayed behind because I wanted to sit by the river alone for a little bit and relax. As soon as they left I grabbed my cedar flute and followed the hiking trail out of camp. I soon spotted a tree that had fallen across the river. It was long enough that it reached from shore to shore and thick enough around that one could walk all the way across the river on it. I walked out to the middle and sat down and began playing. It was so nice being able to sit over the top of the water like that and just watch it flow under the tree while I was playing.

It was so peaceful that I never wanted it to end, but all too soon the sun began to set and I wanted to head back to camp to get a fire going for us. Before I left I called out to the forest and said that we were camped here for the weekend and I extended an invitation to the sasquatch to come and visit if they would like to. I didn't know whether anyone was there listening to me, but it felt like the thing to do at the time. I walked back to camp and tried to get a fire going but this was easier said than done because the pit was full of water and without dry paper and tinder I was having the hardest time getting a fire to catch. I'm ashamed to admit that I never did get a fire started that night so we retired early. Now I carry fire starter in my backpack so I'm prepared for nights like this.

The next morning when we awoke, we began getting ready to return to the venue for the event. As usual, I was running a bit behind everyone else that morning and I went ahead and washed my face and brushed my teeth quickly right outside of my tent instead of running outside of camp like I usually would have done. Both men were standing there watching me, and waiting for me, and both saw me take my toothbrush, toothpaste, and face wash and place them back into my backpack when I was done using them. I then put my backpack into my tent where it stayed for the entire day.

The Symposium was fantastic that year because suddenly everyone was talking about the kinds of things that pertained to what I was experiencing. One man even spoke of the fact that he had done some experiments with different people who had a habituation situations going on at their house, and found that the sasquatch could and did use telepathy! Another spoke of the possibility of the sasquatch cloaking itself and thereby rendering itself invisible to us. He addressed the fact that some Native American tribes had known of this for centuries! I had a great time at this Symposium and was able to meet other people who were encountering some of the same types of interactions that I was. I also picked up Orbs in some of the pictures that I took there.

Figure 32 Thom at OSS

It began to rain while we were at the Symposium and when we got back to the campsite that evening Charles found that his tent had leaked and was now more of a swimming pool and less of a tent. All of his things were soaking wet and he couldn't possibly sleep in there so I invited him to share my tent with me and gave him a blanket to cover himself up with. When we got into my tent, I got him comfortable on his side and tried to do my best to make sure he was warm for the night.

I sat on my sleeping bag and felt something poking me in the behind. Whatever it was, it was inside my sleeping bag so I pulled back the top portion of the bag and found that it was my toothbrush! This was very surprising to me and I showed it to Charles. I said that I remembered putting this into my backpack and asked Charles if I had put it into my pack or if I had just thrown it into the tent when we left. He said that he distinctly remembered me putting it into my pack, and then throwing my pack into the tent.

I then felt something else at the bottom of my bag down where my feet would be. I reached into the sleeping bag and found my toothpaste way down at the end of the bag. This was beginning to boggle my mind completely. How did this stuff get from inside my back pack down into my sleeping bag? I had mussed the bag up pretty good when I had been reaching down inside of it so I had to smooth it back out and lay it flat again to get inside of it. As I was doing this I found my face wash underneath the sleeping bag!

I showed this to Charles also, and we wondered what the heck was going on and how this stuff got moved while we were gone. After looking around the tent and in my backpack I discovered that nothing was missing, and only these three items had been moved. We discussed the fact that the tent, the sleeping bag and the backpack had all been zipped when we had returned to camp that day so how did these items end up where they were found. Even if I had thrown these items into my tent instead of into my backpack, they wouldn't have landed under and inside my sleeping bag!

I chalked this up to another one of those moments that you just can't figure out and Charles and I began to discuss other things. He was lying down on his side of the tent and I was now sitting up on top of my sleeping bag. We were having a really in-depth conversation at one point when he quit talking in mid-sentence. I distinctly remember the point at which he quit talking also.

He asked me, "So what happened between you and your..."

That was all he said. I thought maybe he was just composing his question still, and I sat there waiting for him to finish the sentence. When he never began talking again I looked over and discovered that he had already fallen asleep. At the time I just thought that he must have been really tired that day, but a few seconds after this I heard the unmistakable sound of heavy footsteps coming across the gravel and into camp.

There were two very different, distinct foot falls coming in to camp and I knew then that our friends were indeed visiting with us that night. I sat wondering if they were the ones who put Charles to sleep and then I heard them walking up to my tent. I had planned to start talking to them and visiting with them but that was the last thought I ever had that night. The next thing I knew it was morning and when I woke up I found that I was half sitting and half lying in my sleeping bag against the side of the tent and I felt very sore.

Charles woke up very soon after I did and I asked him if he wanted to take a walk out into the woods with me so that we wouldn't wake Thom while we were talking. He agreed and we set off up the trail. After walking a bit we sat down on a log and I asked him if he remembered falling asleep in the middle of our conversation the night before. He said he thought we had finished talking. Then I told him about how we had a few visitors walk into camp right after he fell asleep and how I must have fallen asleep right after I heard them walk into camp. I told him how I thought the sasquatch must have put us both to sleep in order to talk to Thom or something.

I could tell that he didn't believe a word of what I was saying because he had that look on his face while I was talking to him. I just told him, "Hear me now, believe me later."

I knew that this wasn't the strangest thing this species has done in my presence. In fact, this is one of the tamer things that they can do to a person. I knew that I couldn't change his mind, nor did I really want to try. I felt that this was something he would have to find out for himself over the course of his research into this species. Everyone has to have this sort of experience over and over again to try and wrap their heads around it or to believe that it can really happen. I just told him that it was probably time to get back to camp and see if Thom was up so we could get to the Symposium to listen to the days speakers.

When we walked back into camp Thom was sitting at the picnic table waiting for us to arrive. I walked up to the table and asked Thom if he had slept well. He proceeded to tell us about the visitors that we had in camp the night before, and how they had spoken to him about a few things. I just looked at Charles and said, "I told you so!"

It made me feel very good to have this validation from Thom about what I had tried to discuss with Charles while we were out in the woods. I don't like it when people think I'm lying about the things I've experienced with the sasquatch, and it was important to me that Charles not think that I was just making this up.

I don't know why they put him to sleep before they walked into camp, but maybe it just wasn't time for him to have this interaction. It seems that the sasquatch will only share themselves when they want to, not when we want them to. I was left with the thought that perhaps they put me to sleep because I wasn't supposed to be part of the conversation that they had with Thom. I did feel a bit left out at that point, but I understood that it wasn't right for me to be awake and they did let me know that they were around and were thinking of me when they went and hid my toiletries from me.

When we arrived at the venue that day, I overheard a few people talking about the night walk they had all taken the night before. I found it humorous that the people who were camped up at the venue went out for a night walk to try to find the sasquatch and never found a thing, yet they had come right into the campground with us to say hello. Just goes to show that if the sasquatch want to spend time watching you, or making contact with you, they will come and find you and if they don't you can walk around all night long and never see anything at all.

This weekend ended all too soon and we made the long drive back up into Washington. I had learned a lot from the speakers that attended this function and I had a lot of information to stew over in my mind. I now knew that other researchers were having these supernatural types of experiences with the sasquatch and that what I was experiencing had been proven to be more common than I had thought. It made me feel better because I knew that I wasn't the crazy lady that others had deemed me to be and now I could stand up and say that these were the sort of things that happened when you had an encounter with this species. If other people hadn't had these experiences, then perhaps the sasquatch just didn't like them enough to allow them into their world. I decided that I would no longer be ashamed to share the strange and wonderful things that I had experienced and I felt blessed that the sasquatch had immediately taken me in and allowed me to experience the wonderful things that they could do.

As I stated earlier, my home life during this time was hell, and as soon as we returned from the OSS I began to have stomach problems. I thought it was an ulcer brought on by my living situation, so I ran back to Clackamas to stay with friends for a while and ease my stress. While I was in Clackamas, I got a call on my cell phone from a woman I will call Jean, that I knew nothing about. Yes, these things really do happen to me!

Jean was a very outspoken woman and she told me that God had spoken to her, and she was to come and pick me up and take me under her wing. She said she was a sasquatch researcher, and that we were going to go camping together and get to know each other. I was about to decline, but camping sounded wonderful and I figured if she was a researcher maybe she could teach me a few things also. We made plans for her to meet me and pick me up.

I really had no idea who this woman was, or what to expect from her, but when she came over I found that she had a great personality and she seemed to be very forthright and honest. I decided that I would go with her and thought we might even have some fun together. She expressed an interest in going to my Washington area. Well, I figured it really wasn't my area anymore and since I hadn't been there in a while and would love to see it again it wouldn't be so bad to share it with Jean, so down the road we went.

When we arrived at the lake I was quite taken aback at the changes that had been made to the area. We could no longer drive in past the lake to get to my favorite camping spot because someone had come in and dug a deep culvert right through the road at the edge of the lake which allowed the water to run off through the culvert. It was far too deep for any vehicle to drive through, and there was absolutely no way to drive around it. We decided to drive on down to the lower lake to check on possible camp spots in that area and I almost wept at the amount of clear cutting that had taken place down there. We never found a place suitable for camping in that area so we drove back to the first lake to try to find a place to camp there.

Figure 33 Our camp at the end of the road

It was starting to get really close to sunset by now and we still hadn't found a place to set up camp. I remembered the spur road that ran up the mountain behind the lake that Cristy and I had walked. When we got to the T intersection, we turned to the left and continued to the top of the ridge where the road ended. This put us up above the lake. There wasn't a lot of room here to set up camp since the road ended in a very small circle nestled between the side of a hill on one side and the tree line on the other. We made do and got camp set up anyway. We both had our own tents, ice chests and backpacks up here. It was fully dark by now and we both retired to our tents for the night.

The next day we left camp and drove into town for supplies. We had spent most of the day gathering what we needed and we were both very eager to get back up on the mountain. When we arrived back in camp I walked over to my cooler to put my food away and she walked over to her tent. When she opened her tent, she instantly backed away from it. She called over to me and asked if I would come to her tent. I was walking over towards her when she asked me to smell the inside of her tent. I couldn't smell anything on the outside so I walked over and stuck my head inside and inhaled. Her tent smelled very strongly of dog poo! It smelled so bad it actually made my eyes water, and I instantly regretted taking such a strong sniff in the first place. I backed out with quickness also! I told her it smelled like dog poo, and she asked me if my tent smelled. I had to walk over and check it out, but the inside of mine smelled just like my tent always smelled.

After I told her that mine smelled just fine, I began putting my groceries away again. I saw her open her ice chest and back away from that also. Even her ice chest had the pungent aroma of dog poo in it! Mine smelled just fine. I know this may be a mean thing to say but I kept thinking to myself that if the sasquatch don't like you, there must be a reason why I shouldn't like you. I remembered how my friend and I had smelled the same thing on the trail the previous summer, and I don't think they liked him at all. Everything she had up on that mountain at that time had been sprayed with that dog poo smell. The only thing I couldn't figure out was how the smell got inside the tent and inside the ice chest, without it being on the outside of these things.

We had backed the car in when we returned from the store and that night we sat in her car looking downhill, down the spur road. She kept saying that she saw eye shine in the trees. I did notice sudden flashes of light that would come and go, but couldn't get a clear reading on them since they were so far away. It seemed that each time she pointed out an area to me the light would be gone by the time I looked where she was pointing.

The next day we wanted to take a hike up and over the mountain to sit by the lake. I thought I had it all mapped out in my head and knew exactly how to get there, but we must have made a wrong turn to the right at some point because we ended up coming out way further to the west than I had planned. But we did have a very enjoyable hike, and we found many sasquatch tracks that day in various sizes. They really get around on this mountain top. Instead of coming out by the lake, we actually came out on the backside of the Bone Field! I was pretty excited to see it, and sure enough there were bones lying all over the place. She climbed up on the huge fallen tree, and I took some pictures of her up there.

I thought I could easily lead us out of the Bone Field and over to the lake, but instead we came out right below the caves that are in this area. I didn't realize we were headed the wrong way at first, but after stopping to breathe and analyze where we were on the mountain, I finally got us down on the correct logging road and headed in the right direction. I had started to panic when I noticed that we were headed into the one area that I never wanted to come to and had, in fact, been warned to stay away from and it was getting very dark underneath the tree canopy. Thankfully this woman helped me to calm down enough to get my wits back again, and I was very happy to finally come out on the main trail and into the sunshine.

While following this main road I noticed that deep culverts had been dug across it all the way up the mountain, not just in the camping area. I wondered why they would want to do this since this road wasn't an easy one to drive even before they dug the culverts. The only traffic that was able to make it up here was a 4 wheel drive with a lot of ground clearance, so it wasn't highly used at all. They had basically made this entire side of the mountain inaccessible to everybody, unless you were hiking in. We stopped to rest at the clear cut and found it full of flowers this time of year.

When we got back to camp, she sat in her car talking on the phone, and I walked the small hillside next to us taking pictures of all the flowers and strange insects that I came upon. When it got dark I took a few pictures around camp and actually caught eye shine! I know that it was eye shine because the moon was still behind the trees to my right and hadn't even fully come up yet and the eye shine is coming from the small hillside to the left of me and the campfire. I had spent the afternoon hiking that hillside taking pictures, and where this light was coming from I found no metal or other reflective surfaces. Could it be an animal of some sort? Yes…

We left here a few days later and went over to a gathering

Figure 34 Campfire and Eyeshine

for sasquatch researchers. Before we left Jean was told by a friend of hers that we should watch for a few sasquatch guests that would be following us over to this new site. We were given things to look for at this new location and we were told to set up camp in the spot where we found these certain clues that we were to be given. I thought this was the kookiest thing I had ever heard, but after the strange things I've been through, who was I to think anything was impossible? When we arrived, we said our hello's to everyone who attended and visited with them for a few minutes before setting out to see if we could find these certain things that we were told to look for. Imagine my surprise when we found a campsite that matched perfectly, and even had a trail that led out of it directly to the lake! This is where we made our camp.

One of the guests at this little get-together was a woman who I have highly respected for years. I highly value the work she has put in for the sasquatch research community. We had been trying to get together and meet, but something always seemed to come up and we had to cancel. She visited with us in our campsite that evening and we filled her in on the possibility of us

Figure 35 Please don't flash me

having two female sasquatches coming here to join us. We showed her all the things that we were supposed to look for... which were all right there in camp. While we were talking about this, she suddenly looks up at me and says, "Do you smell dog poo? I smell dog poo, and I work in a kennel so I know what dog poo smells like!" It was all I could do not to start cracking up and laugh in her face! I mean, really? Had they actually shown up?

I didn't smell anything at the time, but it was decided at this point that we should walk down the trail that led out of this campsite and out to the lake. It was a gorgeous evening, and we were having a great time walking out through the dark without flashlights. We stopped to admire the moon hanging over the lake and the other two began talking. It was such a beautiful sight that I really didn't want to leave the lake, but it was starting to get cold at that point.

As we were walking back towards camp, the other two had a conversation going where one was trying in vain to get an invitation to the others house, and the first was trying her best to redirect the conversation. They stopped to talk with each other about halfway back to camp and I just stood to the side trying not to listen to their private conversation. I was trying to listen to the sounds of the night and see if I could tell if our guests had in fact shown up.

Where we stood, the ground was covered in huge ferns that stretched out in all directions. The tree canopy was keeping the moonlight out so there were a myriad of hiding places all around us. I was just standing there quietly and thinking about how bored I was getting when I heard a quiet voice in my head. I remember thinking it had to be a female voice and it uttered a simple little question that cracked me up entirely. I actually laughed out loud when I heard it. The only thing it said was, "Do they ever stop talking?"

When I started laughing the other two looked over at me and asked what I was laughing about. I told them what the voice had asked me and we all had a good laugh over it. We looked around but couldn't see anything and they didn't do anything else to let us know if they were around or not so we eventually walked back to camp and called it a night. The next day we visited with the rest of the group until it was time to pack up and hit the road.

Figure 36 X found in cemetery

Jean wanted to go home and check on a few things, so we drove north to her place. While we were staying there she took me over to a small cemetery that she knew of in her area. It was late evening when we arrived and we walked through the cemetery looking at the tombstones. At the back of the cemetery we spotted a large X leaning in between two trees. We walked over to check it out and noticed that the branches that made up the X pattern weren't growing there, they had been brought in and the ends had been shoved into the ground. I thought it was completely cool looking and I took a picture of it.

When we were done up north we decided to camp at my favorite lake one last time on our way back to Oregon. When we arrived this time we went ahead and pulled in to the lake and parked at the gravel turnaround. We stayed in a small campsite we found out in the woods. I had been having problems with my stomach this entire summer and was having problems keeping food down. I still thought it was just an ulcer brought on by stress and I was taking "over the counter" medicine for ulcers but it didn't seem to be working. I was getting very hungry by this time, and it was pretty cold out, so we heated up some canned stew we had brought with us to warm our bodies for the night. After we ate dinner, we crawled into our separate tents and bedded down. I just couldn't seem to get comfortable and I was in so much pain that I actually began crying. It didn't seem to matter what I did, or what position I put my body in, the pain just kept getting worse all the time.

At one point I just knew that I was going to be sick, but I didn't have enough energy left to stand so I just crawled out of my tent and over to the nearest tree where I lost my dinner. I was so cold at that point that I decided to crawl out of the forest and into the gravel lot where we had parked the car so that I could sit in it and try to get warm. I made it as far as the gravel area before I once again got sick and lost what was left of my stomach acid. I spent the remainder of that night lying on the ground shivering and being sick. What a way to end the perfect summer camp trip.

When Jean found me the next morning, we took off down the mountain and headed back over to Cristy's place in Clackamas. Since the apartment was now filled with Cristy and the new baby, plus her boyfriend and his friend, I just set my sleeping bag up on the balcony and slept there. A day or so later Cristy found me lying out there disoriented and in considerable pain, and took me to the hospital.

It was determined that my Gall Bladder had died and had actually been rotting inside of me over the summer and I had surgery the next morning to remove it. I don't even remember Cristy finding me on the porch, or taking me to the hospital, so I will be forever grateful to her for that.

I also want to thank Jean, who I spent the summer with for coming to my rescue. Not only did she help to get me out of an unbearable living situation, but I could have possibly died had I been at home alone when my illness came to a head like that. By keeping me under her wing all summer, and then depositing me at Cristy's place, I was able to get the doctor's care I needed. I thank you very much; you truly are one of God's Warrior's!

Chaper 15 – Primal People Conference

After I was released from the hospital I moved to Southern Oregon to be closer to my son. He had been asking me if I would move in with him to help him out for a few months and I finally agreed to it. I didn't want to put so much distance between me and my favorite sasquatch sites, but almost dying had put things into a much better perspective for me. As much as I liked hanging out with the sasquatch, I loved my son more and really did want to spend time with him while I still could.

I moved in October of 2011, and he and I shared a house together until February 2012 when I finally found a place to call my own. Since I had moved into a smaller town with much cheaper rental prices, I was now able to afford a two bedroom townhouse by myself. I was so happy to finally have a place of my own and not have to share it with a roommate anymore.

To my utter surprise, Cristy and her boyfriend also moved to Coos Bay a few months later. I loved this fact because now I had my research partner near me again. It was also nice to have her son Connor around as well. I liked to call him "my little mountain man" and I wanted to be able to introduce him to the forest and teach him about the many things that lived there. There is so much life to be found in just a small patch of the forest and too many people live their entire lives without noticing any of it.

We started taking him to the lake which was right behind my new townhouse, and he loved to just sit on the bank and throw rocks into the water. He named this game "Ploop." When he first started throwing rocks, I began to say Ploop as the rocks hit the water, imitating the sound it made for his benefit. Now when he asks to go down to the water and play, he says, "We go ploop?" It's so cute to hear him ask to go play Ploop in that little baby voice of his. He would sit there all day and never leave if we had the time to let him do so.

We had a very strange event occur when Connor was still quite young. We were in a very large park with a creek that ran through it and ended in a man-made lake with a fountain in the middle of it. It's a really pretty park and it's situated in the midst of some very large trees with different paths running among them. We had walked around the lake on the main trail and then cut off onto a narrow dirt trail that led us under the trees. I headed up a slight incline looking for a place to take care of nature and as I started to step up between two medium sized trees my eyes fell on a very long, very wide, footprint in the loose dirt. Instead of stepping on it, I jumped off to the side so I wouldn't destroy it. I bent down to get a closer look at it to see if I could detect a boot tread or something identifiable in the track. It looked pretty smooth, but it wasn't clear enough for me to tell for sure. I called Cristy to take a look at it.

Since we had Connor in the stroller we couldn't follow them into the brush so we went back down to the trail. As we were walking back down the same dirt trail that we had just come in on, Cristy looked down and spotted a Trillium laying there. We had just passed this spot a few minutes before and there was nothing laying there then. We both found this odd. Then we spotted another dirt path that led up into the brush and we turned to follow it up and look around for more evidence. The only thing we found up there was a guy standing off by himself, doing who knows what. So we turned back around and went back down to the main path.

I couldn't believe that we would find sasquatch tracks in a public park, even if we were outside the normal paths that most of the parks guests strolled down. I knew for a fact that humans do walk up in that area, since we were there ourselves, and we discussed whether or not we thought they could be sasquatch prints. I don't remember the decision we came up with exactly because I myself didn't believe they could be at all. I just knew that the sasquatch couldn't be this close to civilization and not be seen or shot at. At least that's what I thought then.

Thom contacted me that month to tell me that he was putting together a conference that would be held in Richland, Washington during the month of May and he wanted me and Cristy to be sure and attend. It was going to be called the Pacific Northwest Conference on Primal People and after the conference he and a small group of people were going to be renting a cabin in the Blue Mountains for a week. We were both super excited to attend and made plans to do so.

We dropped Connor off to be protected by his aunt and uncle and then we hit the highway for the long drive north. It was really nice to be able to take a long road trip with Cristy again and we had a great time listening to tunes and laughing together. When we finally arrived at Thom's house the night before the conference we found that there were already quite a few people staying there, but Thom was gracious enough to let us crash on his living room floor anyway.

We got up early the next morning and car pooled over to the hotel where the event was being held. Cristy and I had made key chains with glow in the dark footprint casts that included the Primal People logo to sell at the conference, so we set up a table in the corner when we arrived. I'm not much of a talker, or a salesman, so I have to give credit to Cristy here for manning the post all weekend.

There were so many wonderful speakers to listen to during the conference and I met a lot of really nice people. We even went for an outing into the Blue Mountains to an area where Thom often met with his teacher. We must have had a hundred cars convoying up to this area, and we all scattered as soon as we got there. Cristy and I wandered around looking for evidence but all we found was an old elk jawbone so we wandered over and spent the rest of the day listening to a lady playing her drum in the woods.

Figure 37 Cristy (L) and I in Blue Mtns of WA

For me the best part of the conference was that an artist friend of Thom's had brought along a full size bust of a sasquatch head and had also made a life size reproduction of Thom's teacher, Redstripe. I was still coming to grips with how large this species actually was at that point in time and seeing the bust and standing next to the reproduction was a real eye opener for me. Up until that time I just hadn't wrapped my head around how large they really were! For some reason my mind just wanted to think of them as around the size of a large man or perhaps a bit taller.

I think that if I was fully aware of how wide and tall these beings really were, I probably would have panicked more in the beginning when they were so darn close to me all the time. I know to some this isn't going to make much sense, but to me it makes a lot of sense. In order to deal with some of the weirdness that has happened to me during this journey, my mind has had to come to some really quick decisions on how to handle it all. Just checking out has never been an option, I'm far too strong for that to happen. So instead of just losing it, my mind makes it all a little less stressful.

Like with the glowing orb flying in front of me. Instead of going into a panic and losing it or having to deal with the reality that I didn't know if this something could be dangerous to me or not, my mind brought up the glow-in-the dark comets that had adorned my walls. This memory also brought with it a calm peaceful feeling from childhood, so it made it all a lot less scary.

Figure 38 Me with full size reproduction of Redstripe

Having them standing behind me or next to my tent would have been much more disconcerting. Not having a clear idea of how tall and strong they could really be, I was a lot less worried about having my head ripped off and handed back to me. That day at the conference it all became perfectly clear to me. The only thing that I could think of then was how lucky I had been that they really aren't the monsters that some people portray them to be. I'm glad that I've always come to them through respect and love because I wouldn't want to get on one's bad side.

After the conference we drove over to the cabin to spend a few days with some great people. There was still snow on the ground in the shady spots, but the sun was shining and the sky was blue. This cabin had a loft with all the beds located very close together, so instead of being packed into the cabin I opted to sleep in my tent outside and Cristy decided to join me out there. We set up the tent outside the cabin while everyone else was inside picking their sleeping quarters.

The first night we were there it seemed that cliques were formed, and Cristy and I both took offense at something that was said to us that evening, so we decided that she and I would hike out in the morning and just do what we do in the woods and let the rest of the group be together. When we retired to the tent that night it began to get really cold. I could hear Cristy's teeth chattering for most of the night and I felt so sorry for her. My mummy bag was so warm I kind of felt guilty for having it, but after the night we had spent above the Clackamas River I had made sure to put a Minus Twenty Degree sleeping bag on my Christmas list. The next morning we went into the cabin to see what was going on with everyone else and seemed to receive the same cold shoulder as the night before, so we grabbed our stuff and set off to enjoy the day.

Right outside the cabin was a very large grass area that led down to a creek and we had placed my tent right above that. We walked down to the creek and checked it out a bit and then noticed that there were a few roads running through the area that were used for snowmobiles in the winter. Most of the snow had melted off already and we followed one of these roads along the creek for a while. Almost immediately on leaving the cabin and walking up this road we came across a large sasquatch track. This part of the road was in the shade and still had a lot of melting snow on it which had turned to ice and the track was left in this icy snow melt.

Cristy and I looked down at the track and just smiled to each other. We now knew that our little group was not alone on this mountain by any means. We did nothing but look at the track and continued on our way again. We walked all over that mountainside that day and we saw so many cool sites. As we were walking through a big green meadow, I happened to look off to my right and spotted a really cool sight.

Sitting in the shadows of the large trees was a little brown stump, and the way the shadows were laying across the face of this stump made it look just like a gnome! I sure wish I had thought to grab my camera before we set out, but alas, I had not. We stood admiring it and discussing the fact that simple shadows can truly fool the eyes into thinking you're seeing something entirely different than what's really there.

We climbed up the hill on the far side of this meadow and found ourselves walking through some thick trees. We were having the time of our lives just walking through the snow patches and checking out the new sites. We did find another trackway higher up the mountain and followed it. I looked over at one point and noticed some benches forming a semi-circle in what was probably a clearing at one time, and we walked over to check them out. Turns out that there was an amphitheater out there on the mountain, but it couldn't have been used in some time as there were leaves covering everything and it was starting to rot in places. The benches encircled a small stage area that was now filled with cut branches and wood rounds and other debris, and the tree line began directly behind the last set of benches.

This was a perfect find for us! We couldn't believe our luck that day. Cristy had her flute with her and here was the perfect place to play. I asked her if she would get on the stage and play her flute for me so I could lay back and listen and she was all for it. I started clearing off the stage for her while she looked around for the perfect size log to place her stuff on and we got it all set up. Cristy sat on the stage and started playing her flute while I went from bench to bench trying to find the one that was the most comfortable, and had the best sound coming to it.

I finally found the perfect spot and I lay on my back looking up at the sun coming through the leaves on the trees while I listened to her play. It was the most beautiful thing in the world and I actually got tears in my eyes listening to the hymn she was currently playing. I was wishing that I had a video recorder with me so I could record it all when I heard Cristy falter. She didn't quit playing entirely, she just paused for a second and then it seemed like she couldn't get it back together again. I finally turned my head towards her to see what was going on, but by that time she was taking a breath and leaning her head towards her flute again.

I figured that she had just got caught up in the beauty of the place like I had, and didn't say anything about it at the time. My only thought was that I didn't want it to end so soon. I had been hearing very slight sounds off in the tree line while I was laying there, but it wasn't anything loud or significant and I felt it was probably just squirrels or chipmunks running up and down the trees enjoying the sun as much as we were.

All too soon the sun began fall below the tree line and the air began to turn cold up under the trees so we knew it was time to head back down the mountain to the cabin. Cristy was quieter than normal on the walk back, but I didn't say anything about it and didn't think too much about it at the time.

When we got back to the cabin the others were talking about how they had come across a track that day. It turned out to be the same track that Cristy and I had found when we left that morning. At that point Cristy went outside for a few minutes and then came back in. I was glad that the rest of our party had been able to see the print also, but they were acting like it was supposed to be some kind of secret only they were allowed to know or something. I never did understand that part, but I let it slide.

Then another one of our party came in and whispered really loud in Thom's ear that they had heard a whistle outside and said, "We have company". I couldn't understand why this person would be whispering something like that, especially when they whispered it loud enough to be sure the entire room heard them anyway. To me it just seemed like more of the high school games that were being played while we were there, but I stayed seated to see what would come next. Cristy got up and left the cabin and went outside.

After about a half hour or so, I followed Cristy outside and found her sitting in a camp chair in front of the tent with the hymn book on her lap. She was facing the grass and the creek, with the tree line off to her left and the gravel road to her right. I grabbed my chair and sat next to her knowing she wanted to talk. She told me that earlier she had gone out and left fruit inside the tree line for the sasquatch, and then heard one whistle. So that explained the whistling.

She said she had come outside to play for them, and while she was out there she had the strangest thing happen to her. She was looking down at the hymn book that she had lying open on her lap and she felt the energy in the air change and she swore that this next part happened.

Here is how she relayed it to me... "I was sitting there singing since it had gotten too dark to read the notes and play the flute any longer. I was holding my flashlight in one hand and the hymn book in the other when I heard three footsteps on the gravel to my right side. I then felt something walk up to my chair and lean in real close to me. It was like it was looking directly at me, but there was nothing visible in the light from my flashlight and then it backed off. My face flushed and grew hot but I just focused on my hymn book and I kept thinking to myself, 'Just keep singing.' It was only there for a few seconds. If I had lifted my flashlight in that direction I feel that I would have seen it."

She also told me what had happened at the pavilion earlier when I heard her falter while playing. She said that she had looked up while she was playing and in the tree line behind me she saw a young sasquatch hunkered down with bent knees watching her play. He scooted to the right with his knees still bent and she looked back down to the hymn book. When she looked back up again, it had already stood up and the bottom half of it was gone but she could still see the shape of the top half of it and this looked to be disintegrating into a swarm of bees. He had one arm in the air and that looked like the tail end of the swarm to her eyes. Then she looked back down again, and when she looked back up, it was completely gone. He was a reddish orange color. She stated that the reason she kept looking up and then back down again was to see how many counts to hold a note, so it was "look down two, three. Look up two, three." It was only a few seconds between each look.

I didn't know what to think about this revelation but I was happy that she had finally had a sighting of her own. She asked me not to reveal anything to this group of people because she was still trying to come to grips with everything she had seen. I had to respect her decision on this one so I told her I wouldn't say anything, especially since it wasn't my story to tell. We ended this evening by making a campfire in the fire pit outside the cabin and sharing it with those who chose to come outside and enjoy it with us.

The next morning it was decided that the girls would all take a hike up the mountain. We all set out together but after a short while half of us decided to go back down to the cabin. That left me, Christy, and one other woman continuing up the mountain. We were trying to find the pavilion again, but we were coming in from a different direction this time and thought we weren't going to find it at all. We did eventually run right into it and we sat down to rest and talk. Cristy decided that this was the woman she wanted to share her story with, and she did just that. For some reason she seemed to leave out the part about it disintegrating into the bee swarm, but I figured she just wasn't ready to disclose that part yet.

When this trip ended, we all said our goodbyes and hit the road. Neither Cristy nor I were really ready to get back to civilization yet so we decided to camp for a few more days before picking up Connor. We grabbed the cell phone and looked up free camp spots around the Eugene area where we could go. The first suggestion led us to a really creepy and spooky place that we both decided was not where we were going to camp, so we headed to our second destination. It was pretty far up a mountain road and it was starting to get dark by the time we found the correct road. When we pulled into this unimproved camp site it was fully dark and we were the only people in the area. We hadn't seen a car for quite some time, and there were no cars parked in this area. We backed up next to a camp spot and turned off the truck.

As we were getting ready to get out of the truck, it suddenly bounced down and then back up with a very powerful jolt. It felt just like something had jumped into the bed of the truck, which would be impossible because it was full of our camping gear and covered with a tarp. We looked at each other and both said, "What was that?" Since it was clear that neither one of us had the answer, I told Cristy to turn the lights on and step on the brake so we could light up the front and back of the truck. I turned around and looked out the back window and didn't see anything on the tarp and it didn't look rumpled up as if something had jumped on it either. Of course we couldn't see the ground behind the truck or on either side of it.

I turned toward the front again to tell Cristy that I hadn't seen anything in the back of the truck when I suddenly heard what sounded to me like someone sprinkling pine needles on the cab of the truck. Cristy said it sounded like claws sliding across the top of the truck. We didn't agree on what made the sound, but we both agreed that neither one of us was getting out to see what it really was. I kept expecting to see something jump off the cab and onto the hood of the truck, but nothing ever did. We never felt the truck move again, and we never heard another sound either.

We decided we would just sleep in the truck until morning when we could check it out in the daylight. We just sat in the truck trying to stare out the windows into a pitch black night which is all but impossible. At one point I turned towards the window and caught my own reflection which really panicked me because I thought someone else was looking in at us. I couldn't lay my seat back fully because there was too much gear piled behind it, so I had to lay with my head much closer to that window than I wanted to.

When the sun rose the next morning, we literally fell out of the truck because our limbs were so stiff. The first thing I did was to look up at the cab of the truck expecting to see either pine needles or claw marks scratched into the top of it. It hadn't been windy that night, so if it had been needles there should have still been some up there. Since we were parked on a gravel surface we didn't notice any tracks around the truck anywhere either. We did try to recreate the up and down bounce that we had experienced in the truck and we found that to get that same effect we would have to either jump on the back bumper or push down really hard on the back tailgate. Anything else would rock the truck forward and then back, and that's not the motion we felt. We never did solve this mystery and just let it go.

We set up camp in a grass covered area and found a piece of burnt wood lying near the fire pit. It was in the exact shape of a very large foot and I thought it was so cool that I kept it.

This was a very beautiful area nestled beside a river with numerous waterfalls cascading down off the mountain. We stayed at this site for a few days, but never did have any night time activity and never saw any wildlife wandering around anywhere. So whatever it was that bounced down on the truck our first night there remains a mystery to this day.

Chapter 16 – Taking the Baby to Meet the Clan

When Connor turned one year old, we felt he was ready for his first camping trip and we wanted to take him back up into Washington to meet the clan. Cristy and I were both very anxious to get back up there and we knew we would be safe even with a small baby in tow. The clan had never previously let any bear come around when they were in the area, and even though we had spotted cougar tracks in an adjacent ravine, none were ever seen on that side of the mountain so we knew without a doubt that nothing bad would happen while we were up there.

We left on a cold and damp morning in a pickup truck so completely loaded we couldn't even see out the back windows. We knew that bringing a baby along on our two week stay on the mountain would mean including more stuff than we usually carried with us, but I don't think we were fully prepared for the total amount. We eventually fit it all in, got the tarp tied down, and hit the highway for our seven hour drive up the coast. By this time, neither one of us were really feeling the excitement yet. Luckily, about half way there, the clouds parted, the sun came out and lifted our spirits which got us back in the camping mood.

We reached our destination a couple hours before sundown so we still had plenty of daylight left to set up camp. After securing Connor in his mobile play pen, we got busy with the task of emptying out the truck and setting up the camp site. We quickly realized that trying to stick to our usual mode of doing things in camp was going to be impossible with a toddler demanding attention all the time so Cristy kept him occupied while I finished getting set up and started the campfire.

We had picked up a fully cooked fried chicken on the way to camp so we ate that the first night instead of trying to prepare a meal over the campfire. A little boy walked into camp while we were eating and advised us that he was camping up there with his dad and two brothers. He was such a little cutie we allowed him to talk our ear off for a while. He told us that his dad had come by earlier and taken all the firewood out of our campsite and put it in theirs. He even apologized for it so we told him that it was alright because we had brought some wood up with us, and he eventually wandered back over to their campsite across the lake.

We were pretty tired after the long drive up and we went into our tents early after darkness fell. I got into my sleeping bag and laid there just listening to the sounds of the night and thinking about how elated I was to be back here at my favorite lake again. I was wondering if my friends were back up on the mountain and if we would find some evidence of them when we went hiking the next day when I heard a sound in camp. It sounded just like plastic scraping across gravel. I've heard that sound a million times while camping... it was the sound an ice chest makes when it is being pulled across gravel! I knew that's what the sound was, because not only have I heard it so many times in my life, I felt thirsty upon hearing it and could taste a cold soda in my mouth. Talk about subliminal messages! I quietly lay there listening for any other sounds and then I heard it again. Plastic being pushed across gravel! I had to smile at that one. Perhaps he/she had tripped over the ice chest and then had to put it back where it was before. Or maybe they just wanted to know what was in it.

I was lying in my sleeping bag listening to see what they would do next when I felt the all too familiar feeling of hands being placed on either side of my head. Then the constant pressure being placed on my temples again, like someone was trying to squeeze it with their fingertips. I just lay there feeling this sensation of pressure and warmth for a few seconds or so then I lifted my head off the pillow and shook it. The pressure released, so I lay back down and once again felt the hands wrap around my head and start to squeeze my temples. I spoke harshly and said, "Just don't start squeezing my head together like they did before". The pressure started easing up until it was gone, and I rolled over on my left side facing the tent wall. As I was dozing off I thought I heard two steps across the gravel coming into camp, but I must have dismissed it and fell asleep.

The next morning when I got out of my tent Cristy snapped at me "Why did you pee outside my tent?" She continued to say that if I didn't want to walk all the way over to the bathroom, why hadn't I just peed over in the bushes behind my tent instead of peeing behind her tent! I told her that I had never left my tent after I zipped it up the night before and I hadn't used the bathroom all night! She then explained to me that she had heard two steps on the gravel coming into camp towards her tent, and then heard the unmistakable sound of someone peeing right behind her tent. She described the sound as a stream of water that splashed on the ground. She said she knew what it sounded like when a person was peeing outside, and I believed that.

We looked around to see if we could spot a wet area near her tent, but everything was dry already since this was now about ten am. We looked around to see if we could figure out which way this "something" had entered camp, and did find a path that led into the woods behind her tent. The ground was entirely moss covered almost all the way up to her tent. That would explain how someone, or something, could get into our campsite pretty quietly and would only be crunching the gravel once they were already on top of us. This trail was lined with bushes and relatively thin, but opened up once you were in the trees. It would be quite easy for our tall friends to come in and say hello and not be heard walking.

After looking for clues to our unknown visitor, we drank our coffee while Cristy fed Connor his breakfast. We then put Connor's backpack on him, and walked up the mountain via the old logging road. About halfway up, Cristy excitedly pointed out something on the side of the road. It turned out to be a very small sasquatch print. It was walking straight across the road, and then veered off to the left. There was a larger juvenile print in evidence also. It turned in the direction of the little print, and the little prints ended. The larger of the two then continued off onto the side of the road. The small print was deeply embedded in what was now dried mud and it looked like it had stepped off an area of high rocky ground and down into a lower area which probably had water in it at the time. The larger one was not very distinct, because it had stayed up higher towards the dry side of the road.

The story these prints told was that these two were crossing the logging road and the little one had tried to wander off in a different direction, but the older one had obviously picked up the little one and then continued into the vegetation on the side of the road and off into the tree line where we could see the trackway in the mud there. It looked like the younger one ventured over and played in the water of the mud puddle, and the older one had picked

Figure 39 Baby sasquatch print

it up and continued on its way into the trees. I took pictures, and Cristy made a video. Then we destroyed the prints.

When we finished our hike and got back to camp, the little boy came back over and advised us that his father had a gun in his camp. He said that his father was afraid of bears and he would shoot whatever came into their camp. I wasn't sure why he was telling us this but I made a mental note of it and after he left, Cristy and I talked about this. We knew who also lived on this mountain and we didn't want one of them to be shot at by mistake.

That night after full dark, Cristy and I sat by the fire talking and mentioned how sad it was that the frogs weren't sounding off around the lake like they used to. We thought that maybe since the culvert was put in the frogs had followed the running water on down to the other lake or something. I sure missed hearing them at night. I eventually said that I was going to go into my tent, and Cristy said that she was going to sit out by the fire for a while longer before she retired. I got into my tent and put on my sweats and sat on my sleeping bag. I turned on my reading light and grabbed my book and read for a few minutes while listening to the fire popping outside.

I decided that I didn't want to read... I just wanted to enjoy the night, so I turned off my light and sat on my bag in the dark. It was at that time that I heard Cristy's tent unzip, and then zip back up, and I remember saying to myself that Cristy just went to bed. Not that I was worried about it, I just like to know where everyone is when we're out in the woods. I laid there trying to decide if I was really tired enough to go to sleep, or if I wanted to smoke another cigarette first. I decided on the cigarette and just as I sat up to reach for a smoke, I heard loud footsteps walking across the gravel over by the fire pit. I then heard the fire pop again, but this was like when you poke it with a stick and it pops and the wood shuffles a bit. That surprised the heck out of me because I thought that Cristy had already gone to bed. So I thought then that she must have just put something into her tent, or checked on the baby, and then stayed outside.

I was groping through my backpack in the dark when I heard the footsteps walk up to where my tent was pitched. I quit moving to hear if she was going to say something to me. I didn't hear anything for a few more minutes when the footsteps just walked back into camp again heading towards where we had set up our cooking area. Now I was really wondering what she was doing out there. I was having problems finding what I was groping for in the dark, and I knew then that I had left my smokes out in my camp chair by the fire. I figured if Cristy was out there, she could get them for me, and I was reaching up to grab the zipper to my tent when I decided I didn't really need a smoke anyway and put my hand back down.

I was sitting in the dark listening to Cristy walking all around our camp with these really heavy footfalls, I was thinking about how she was going to wake up the baby, he was going to start crying and I once again reached for the zipper to my tent. I was going to lean out and tell her to be quiet before she woke him up, but, once again, I decided not to unzip the tent. It was her child after all, and I was sure she knew how much noise she could make before he woke up. It would be my luck that my yelling out to her would be what ultimately woke him up.

I laid back on my sleeping bag and was just getting comfortable when I heard the sound of metal clanking together on the picnic table and then the ice chest scuff. My first thought was, "Oh, she's getting a snack. I could eat"! But once again, I talked myself out of getting up, unzipping my tent and going outside to see what Cristy was up to. I started to get sleepy listening to the sounds of her wandering all around camp. I suddenly got a memory and a visual in my head of the times I would go camping with my family and my mom would tuck us kids into our sleeping bags and then proceed to clean up camp and put all the food away. I got such a warm cozy feeling from that memory that I rolled over facing the side of the tent and closed my eyes and let the sounds, and the memories, lull me into a restful sleep.

When I woke up, the sun was shining brightly and I heard Cristy outside making coffee. When I climbed out of my tent, she looked over at me and asked if I had heard the sasquatch walking all around in camp last night! I just looked over at her and said that I thought it was her in camp last night! I told her that I had thought that she just put something into her tent and stayed outside. She told me that she had gotten into her tent, and had seen me turn off my light, but had never gotten back out of her tent.

She said that as soon as she had zipped up her tent and gotten in bed, she heard two of them walk into camp. They then proceeded to walk around our campsite and check everything out. I busted out with laughter and couldn't stop! I thought of Cristy in her tent with her new baby, hoping she wasn't scared to death during the entire ordeal. I thought of little Connor, who luckily slept through the entire thing... and I thought of me… the brave researcher who sat in her tent throughout the entire episode and never once unzipped it or even peaked out the window to see who was really making all that noise.

We noticed the scuff marks in the dirt and gravel in front of the ice chest, which proved that the ice chest had in fact been moved. The only other thing which we could determine had been moved was a nail. It was a rusted nail that Cristy had earlier spent time trying to get out of the gap between two of the boards that made up the top of the picnic table. It had defied her attempts to release it from its prison between the two boards, but the sasquatch was having none of that! The nail was calmly lying on top of the picnic table. Perhaps they thought that Cristy had needed it for something?

They did make a lot of noise in camp that night too. They were not trying to hide the fact that they were there at all… not at all stealthy or quiet. I'm left to wonder if they really did want us to come out of the tent to say hello.

The little boy and his family were packing up to go home. We know this because he made a point of coming over to tell us and say goodbye. He was such a sweet little boy and a fountain of knowledge. We never heard his father's gun go off, so I can only assume the sasquatch had left them alone.

Cristy and I decided to take a hike up the mountain again. This time we went all the way up to the top and tried to find the rock wall that I had found quite accidentally the last summer. We never did find it, but we walked to our favorite clearing area and stopped for lunch. We took our shoes off and let Connor walk around for

Figure 40 Cristy and Connor

a time, then ate an apple and had some water. We wanted to go down into the bone field again, so we set off after getting Connor back into his pack. Cristy was having the time of her life hiking with her new little boy and he was having the time of his life being out in nature with his momma. As we walked down the ravine towards the bone field though, we came upon some very fresh bear scat. Steaming fresh, in fact, so we made a U-turn and left the way we came. We didn't want to start anything with a full grown bear.

We decided that we had better wander back down the mountain and go collect firewood before it got too dark. The clouds were rolling in, and they looked like rain clouds. When we got back to our campsite, we found that we had some new campers set up near us and they were staying for the weekend. They came over and told us that some nasty weather was predicted, and that God had told them that we needed firewood! Imagine our surprise when they backed their truck into our campsite and unloaded a bunch of wood for us! I didn't know what to say, and felt that thank you just didn't say enough. We all agreed that God is great and we thanked them profusely and asked what we could do to return the favor. They told us that our thanks were good enough for them. I will always remember this very kind couple.

The bad weather did come, and the wood was greatly appreciated! That night as we sat around the fire huddled in our coats, we heard a lot of footsteps in the forest around us. There were sticks cracking in the trees, and rustling in the bushes that bordered camp. We were doing great about not shining the light at every noise we heard and didn’t feel threatened in any way. Then Cristy turned to me and said "Did you hear that?"

I had heard that, in fact. It was the sound of two steps, then a stick crack, and then the vegetation settling.

Her eyes were huge, and I got up out of my chair and

stepped away from the firelight so I could see better in the dark. Something large had walked up from out of the woods and sat in the bushes a few feet from her on her left side. It was sitting just beyond the firelight by the picnic table. I just got a big smile on my face and gave my respect sign to it, and sat back down in my chair. She looked at me funny, but I just looked away with a smile on my face. I didn't want to scare her, because this thing was really, really close to her. But I knew it didn't want to hurt us. If they wanted to hurt us, they could have had us the night before, or any of the other nights we spent in this area in the last four years.

We just continued to sit by the fire feeling elation, and yes, a touch of the "what if's" and let it spend time with us. What a wonderful experience! For four years now we had invited this clan to come on in and spend time around our campfire and now one had finally accepted our invitation! I don't know for sure which one it was and I wasn't about to walk up to it to find out. I didn't want to scare her/him off, or to seem aggressive about it being there in any way. I'm not sure exactly how long it stayed with us, but it was quite a long time.

The next morning the sun came out and showed us just how dirty we were, so we decided to have a spa day. We heated up a pan of water on the breakfast fire and got out the cleanser and the toner and the moisturizers, along with our last hand towel and some cotton pads. We were somewhat hesitant about using the hand towel and ripping it into washrags and not saving it for dishes later, but we also knew how good it was going to feel scrubbing our skin with good smelling soap and hot water. We weighed the pros and cons, and then decided to just go for it and use them anyway. We had a great time washing up, and laughed and joked around while Connor ran in circles around camp picking up sticks, leaves, fir cones, and rocks to play with. His favorite toys the entire time we were on the mountain were sticks, rocks and leaves.

Cristy finished up and started cleaning Connor, so I walked down to the dock with my washrag. I stopped at the edge of the dock and took my shoes off, then walked out to the end where I could sit and enjoy the view. I was using the warm cloth to wipe my arms and legs when I looked down and caught sight of my feet. Man, they were filthy! So I used the wash cloth to clean my feet for a while, and had to dip it into the lake at one point to get it wet again. I located a small twig and used that to get the dirt out from under my nails. Then I lifted one leg up, and brought my foot to my face, so I could check for cuts and abrasions. When I caught sight of myself in my reflection on the lake it was so funny I couldn't help but bust out laughing at myself. Sitting on a lake, at the top of a mountain, cleaning myself like a monkey with a stick! It was at that point that I thought to myself, "Oh well, at least we're alone up here" and I heard a voice in my head say, "No you're not." I laughed so loud at that I nearly fell into the lake.

One of the things we were talking about while having our "Spa" treatment was that we now had to go gather more firewood. All we had left were some wet pieces that didn't want to do anything but sit and sizzle, and it was going to be another cold night. Cristy began to get the baby bundled up and put him into the truck while I turned and placed my paperback book onto the picnic table. I then went to get my little three by five journal out of my camp chair just in case it started to drizzle while we were gone and walked to the picnic table with it. I opened the cover of the paperback book and placed the journal inside it. There was the tarp overhead to help keep it dry, but in case of wind I wanted to put something on top of it so it wouldn't fly across the camp site. I put a bright red pot holder on top of the book thinking that would keep it in place plus soak up any drizzle that might get under the tarp. I then gathered together all the little bottles of facial things we were using and put them in the middle of the table, and placed the unused cotton pads next to them.

We jumped into the truck and drove out to the road to collect firewood from some of the big piles of wood that the loggers had left behind. This wood has been sitting around this particular landing for three years or more and is perfectly seasoned fire wood. It took next to no time to throw together a truck load, and we were back in camp approximately twenty minutes later. Another good thing about these large firewood piles is that they were right at the end of the road that we camp on. No one can come in without us seeing them first. That way we don't have to worry about the wrong people being near our camp site while we are gone.

We drove back up the road and backed the truck into camp, and began to throw the wood out. When we were done wood tossing, I stayed in camp sorting and stacking it, while Cristy drove the truck out to turn it around. I noticed her leaving the campsite at a steady pace, but when she got to where she could see the dock out of the driver's window she stopped. I heard the truck door open and I looked up to see what was going on. She was bolting down towards the dock and I wondered for a fraction of a second what she was doing. I went back to doing

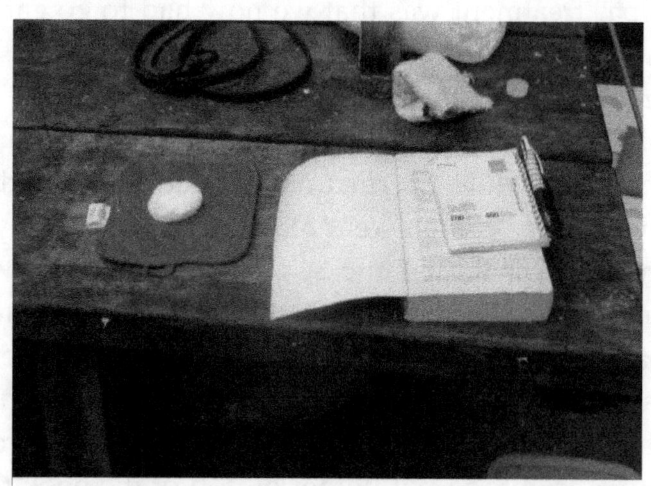

Figure 41 Book and potholder on table

what I was doing. She got back into the truck and continued on to park in our parking space. I dismissed it and figured she saw something cool and went and got it.

As she was parking the truck and getting Connor out, I walked over to the picnic table to get my water bottle and I noticed that my book was open, and my journal had been flipped around. My first thought, of course, was that Cristy must have opened my book before we left. But then I remembered that she was already in the truck when I got there, so she couldn't have. I started looking closer and paying attention to the entire scene and couldn't believe my eyes. Not only was the cover of the book open, and the journal moved to the side (and flipped over), but the pot holder was placed off to the left side, and now had a cotton pad sitting right in the middle of it. That really caught my eye. A bright red pot holder, with a bright white cotton pad was lying right in the middle of it. Nothing else on the table looked like it had been moved.

As I stood there puzzling over this, Cristy walked up behind me, and I asked her if she had left the book open like that. She replied that she hadn’t touched it. She mentioned that perhaps the wind had blown in and done it, and I told her that we hadn't felt any wind that day, but perhaps. So she put the pot holder back on the book and tried to mimic wind. First she tried blowing on the side of it, and then by just hitting the book on the side. Nothing we did moved the pot holder to the side of the book where it had been laying. After she tried that for few moments, I meekly said, "But what about the cotton pad? Did the wind move that onto the pot holder after the wind blew it off the book?" We had both forgotten about the cotton pad while testing our wind theory. It had been sitting in the middle of the picnic table with the rest of our spa day supplies when we drove out of camp.

She told me that this could wait because she had seen something strange down by the dock when she was backing up the truck, and I needed to come down and see this one. We walked down the little path out of camp, and as soon as we got to the road leading to the dock, I looked up the road and could see two medium sized white things lying down on the ground. I looked over at Cristy with a puzzled expression, and she told me that they were towels. Two white hand towels!

Figure 42 Towels lying in road

I couldn't believe what I was hearing, or seeing for that matter. They sure as heck hadn't been down there earlier when I was sitting on the edge of the dock cleaning my feet! There were a few tire tracks in the gravel from trucks that had come in the day before to fish the lake, but no new tracks were there. So that smashed my first theory, which was that they must have fallen out of someone's vehicle. The position they were laying in on the road wasn't far enough apart to have fallen out of both sides of a vehicle, and besides that we hadn't seen anyone driving in when we were collecting firewood!

What we were looking at were two clean white hand towels, lying apart from each other on the ground. They were bunched up. While we were walking up to them, I took a few pictures to show exactly where they were laying, and the tire tracks still showing on the dirt roadway that led to the dock. When we came up to them, I noticed that they were very clean white towels and the only dirt on them appeared to be hand prints that were left from dirty hands. I leaned over them to take photos and the hand prints were very clear. While Cristy held Connor back away from them, I lifted up the first white towel and found a large rock lying under it. The rock was egg shaped, and actually looked like a hard boiled egg that has had the shell portion of the egg peeled back on one side. A very unique rock indeed. So unique in fact, that Cristy immediately recognized it as one of the rocks from the fire pit in our campsite! As we were looking at it, I did notice a char spot on one side of the rock.

We walked over the few paces to check out the other towel and see if there was anything under this one. This towel also had a dirty hand print on the outside of it, but there was nothing inside of this one. We took plenty of pictures of the towels and then took them into camp with us. Cristy immediately walked over to the fire pit to show me where that egg shaped rock had been lodged ever since we first arrived, and yes indeed, the rock that had been there was gone. All that was left was a hole where a rock should be laying. No surprise here but the one in her hand fit into that hole perfectly!

Figure 43 Rock under towel

I'm sure the sasquatch clan left the towels for us to clean up with, but I have never used my towel and it remains in my room today with the rock still inside of it. I don't believe that Cristy has ever used her towel either and neither one of us will ever wash them.

A friend I shall call Eve and her son, Carl, drove up the next day to spend some time on the mountain with us. They had camped up here a lot and had some great experiences with the sasquatch. She proceeded to tell me a story about a time they were camping and went out hiking and came back to camp to find a present had been left for them. They had been searching for the cap to their butane bottle before they left camp and had spent a bit of time in the search before deciding to just go hiking and look for it later. When they arrived back in camp they actually found that a butane bottle cap had been left for them and they wondered where it had come from. It wasn't the same color as the cap on their bottles, so they knew it wasn't one of theirs. So the sasquatch have been known to leave a person something they need! This was a great confirmation to me and Cristy and we knew that we were correct about where the towels had come from.

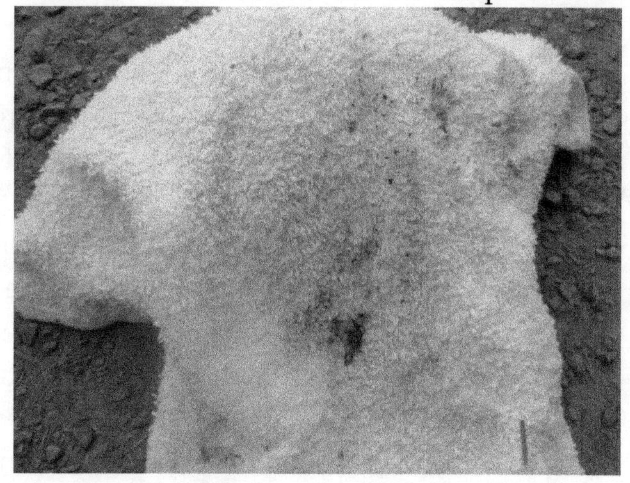

Figure 44 Dirty handprint on towel

They set up camp in the spot next to ours and we proceeded to spend some quality time around the campfire together. Carl played his guitar and then his drums for us and we all sang and danced and had a fantastic time. He's a talented musician and we all enjoyed ourselves very much. We never heard a thing from the sasquatch that night, but I'd like to think they were hanging out listening to the good music he was playing. Eve and her son had had a great interaction with the clan in this area previously and they told us all about what happened. It was decided that we would all head out the next morning together, and they would show us where it happened

When we headed out the next morning, Cristy elected to stay in camp with Connor, so the three of us went without her. A short ways up the hill Carl wanted to show me a spot where he thought the sasquatch sat and watched him one time, so we left the road and headed into the trees. The trees opened up and I noticed that we were now standing at the bottom of a hillside with a wide muddy area directly in front of us. A logging road that hadn't been used in quite some time came down this hill and then turned off to the left of us and continued on.

The second thing I noticed here was that there were elk prints everywhere! The prints came down the hill and bottled up at this precise area where the vegetation created a wall of sorts. We could see that the elk must have been prancing around by all the quick turns they were doing. It really did imply bedlam on the part of the elk. Off to the sides of this area and off to the sides of the elk prints, we found sasquatch prints! It sure showed us how easy it would be for the sasquatch to herd them down into this little cove like area, and how easy it would be to pick one or two for dinner.

He had other things to show me so we headed back to the main road and continued walking up the mountain. We came across some small branch pieces left in the road in the form of a P with a really nice tree whacking stick lying next to it. We had been in search of one for him to own at the time and I bent down to pick up it up for him and found that it was actually tied to the ground with two pieces of Sorrel that were growing there on the side of the road. They actually braided the Sorrel together to hold the stick down. It was a very curious find.

We came to an old logging road that led into the trees and I recognized it as one that Cristy and I had previously walked. It entered the trees a short way in and is a nice cool place on hot summer days. Carl and I walked a short way up this road and saw that we were following one set of elk tracks that just seemed to be meandering through but then began prancing wildly around in the mud. I remember the prancing explicitly because he said that it reminded him of the movie Rudolph the Red Nosed Reindeer when Rudolph gets a kiss from a doe and then prances around saying, "She likes me, She likes me." He even did the dance for me out there in the woods and we both had a good laugh over it. I don't think the elk was laughing in here because the tracks led into this area but they never led back out, and we found two very large sasquatch tracks in the mud here with the elk tracks.

One of the sasquatch prints was up on the side of the road, and the other one was in the mud next to the side of the road. It seems that the sasquatch was trying to stay in the tree line as much as possible, but had to step out into the mud in order to get to the elk. I took a picture of the sasquatch print with his foot in it but didn't have a tape measure with me to measure it accurately. He told me that he wore a size eleven shoe and this track is a lot bigger

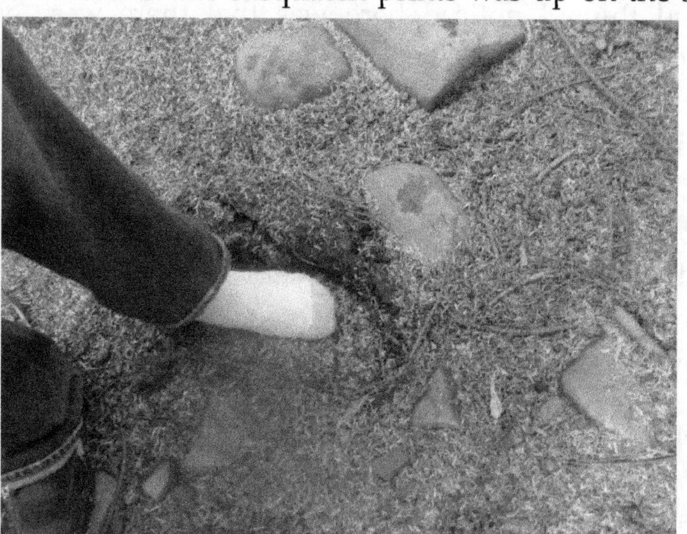

Figure 45 Footprint in mud near elk tracks

than his foot. I tend to think the sasquatch had elk dinner that night.

We then found what looked to both of us to be a doorway under a very lush tree. We walked over to it and pushed back the vegetation and saw what could only be described as a hallway. It was about four feet wide, give or take a foot, and the canopy covered it totally. We entered this "hallway" and followed it directly to where we saw the elk tracks milling about at the bottom of the hill! So this is how the sasquatch could get from one spot to the other without leaving a bunch of tracks in the muddy trails. They sure do know how to stay hidden. I think we were both pretty impressed with this find. I know I was.

We then walked down into the forest on the other side of the main road. Carl wanted to show me where he was standing when he started to hear tree knocks around him one day and then a tree fell very close to him. As we were walking through here I happened to look over and at the edge of the timber, where it meets with a clear cut, was a tree shaped like an arch. Now, I know that the standing timber can take a beating from the wind at the edge of a clear cut, but this intrigued me for some reason. I decided to take a detour and go on over and check it out. I found that it was a fairly fresh tree that had been bent over, and the top of the tree had been pushed under an old fallen tree. Not just pushed under the old tree either. The old tree had to have been picked up slightly to get the top of the fresh tree underneath it as far as this one was pushed under. The tip of the arched tree actually came out on the other side of the dead tree.

It was starting to get late by this time and we headed back to camp to tell Cristy about everything we had found. She was very regretful that she hadn't come with us and I was sorry that she had missed it all. We had dinner and then found ourselves around the campfire being serenaded by Carl once again. He started to play his drum that night and for some reason the beat just kept getting faster and faster, and louder and louder. There must have been something inside him at the time that needed some release that night. It didn't sound bad to me but it was getting way too loud to sit next to him anymore, so I stood up and walked away from the fire.

I was standing off by myself in the darkness when Eve came and stood next to me. We said a few words about how loud he was playing right then and I looked off into the tree line. I noticed a bright white light in the woods off to my left side. As I began to turn my head to get a better look at it, it suddenly moved very rapidly to the west until it was directly in front of Eve and me. It then stopped and hovered there. It was about eye level with us and outside the tree line about thirty feet or so away from us... not far at all. I couldn't take my eyes off of it! I was so thrilled to see the orb again. Perhaps it was a different orb since this one didn't have a tail attached to it. It was just a glowing ball hovering there in the air in front of us.

It just hovered there for a while and I felt like it was watching us. It wasn't casting any light off it.. at least not enough to light up any of the vegetation under it or behind it. It then bobbed in an upward direction once and took off towards the west as rapidly as it had approached. When I finally lost sight of it through the trees I remembered that my friend was standing right beside me and I turned to look at her to see if she had just seen what I had just seen. I knew no one was going to believe that I had just seen another orb with the naked eye and wanted to know that I had a witness this time. She said that she sure had seen it and I was so happy that I had been able to share this sighting with someone else. There was no way that it could have been a reflection because it was emitting its own light out there in the darkness. I felt so blessed to have been able to see it again and I said a silent thank you to the Universe for allowing me to see something this amazing more than once in my lifetime.

We asked the others if they had seen anything, but Carl said he was looking at his drum and Cristy said that she was looking into the fire. I'm so sorry that they missed it because it truly was an amazing sight. The rest of the night passed without a sound from the woods. Eve and Carl left early the next day since she had to get back to work. Cristy and I still had a few days of vacation left and decided to stay around even though it looked like the weather was going to turn foul again. Neither one of us ever wanted to leave when we were here.

The next morning I left camp early and walked to the little stream that flows through this area. I sat on the bank and drank my coffee and reflected on all the things I had witnessed this trip and tried to find answers for some of the stranger ones. It seems like I'm always trying to wrap my head around another mystery when I'm out in this area, but I wouldn't have it any other way, to be honest. I decided that I wanted to walk over to the secluded campsite where I had often camped alone. When I got there I found that the entire circle was bathed in sunlight and I went over and laid down on the moss and sorrel to warm up a bit. I was feeling very melancholy because I missed camping in this spot and because I knew that we had to leave soon.

I spent some time communing with my favorite old rock, listening to my favorite babbling brook and talking to the birds that kept coming in and singing for me. I then heard the muffled sound of a stick breaking into the forest floor. I'm sure all know the sound that I'm describing. Instead of the loud "snap" of a large limb breaking off or snapping off into the air, this is the sound of the twig or branch being broken into the ground. I just got a great big grin on my face and said "Hello, my old friend."

I then heard the loud snap of a twig breaking. I guess they decided not to try to muffle their sounds since I already knew they were there. I heard the unmistakable sound of something big sitting down in the brush, and I had the most irresistible urge to go grab my tent and stay right there in my favorite place in the entire world that night.

But I knew that Cristy would want me in camp with her and we had to return home the next day. Instead, I just lay there and cried for a few minutes and wished that the world didn't always have to move on and everything could just stay the same forever and ever. It made me feel good inside that they had found me in my old camping spot. It also made me very sad that I couldn't just come up here and camp by myself whenever I wanted, or needed to, anymore. Many times through the year I do feel the need to be alone up there and to spend some time with the clan. Even to this day, it's a need that eats me up from the inside out. I miss them, and I miss my magical spot constantly.

I spent the remainder of our time together just lying in the moss and the sorrel and enjoying the company of an old friend. I talked and they said nothing. I sang and they said nothing. I laughed and they said nothing. And yet, it was one of the best days I had that camping trip. Eventually, I heard it stand up and walk off into the woods and I made my way back to our campsite.

Later that day, the clouds blotted the sun from the sky and it began to rain in heavy sheets that never did let up. We left early the next morning and began the long drive back to Oregon. Cristy and I both had tears in our eyes as we drove down the mountain that day. We both knew it would be quite some time before we could come back up here to visit again.

Chapter 17 – Oregon Clan

After we got back to Oregon, Cristy and I agreed that it just wasn't prudent to make the long trip up into Washington with a baby more than once a year. It hurt to admit it but we were going to have to say goodbye to the clan that was there and seek a new camping spot somewhere closer to home. I just hoped that we could find a new clan who were as interested in making contact with humans as the Washington clan was and who would treat us as kindly as they did. I wasn't prepared to lay bets on whether this would happen after the unkind treatment we were given up above the Clackamas River. It became apparent to me then that not all sasquatch wanted humans hanging around their territories.

I spent the next summer camping/gold panning on a major river here in southern Oregon with Cristy's father in law. I knew that he wasn't the kind of person who would make the sasquatch feel welcome in our campsite, but it gave me a chance to explore the woods around that area and see if I could locate any signs. We spent most of the summer out there and I never did see any sign of them or hear anything out of the ordinary. By this time I had pretty much given up on finding a new clan to spend time with. Now I was just camping for the love of being outdoors and hiking up the rivers we camped next to.

During that summer I heard that a store near me was having a "Bigfoot Day" and they advertised that an experienced bigfoot researcher was going to be in attendance to answer questions and give a presentation. That really got my attention. I hadn't met anyone else in the Coos Bay area that was even remotely interested in sasquatch and I was looking forward to meeting this fellow researcher and hoped that he/she lived in this area and would be up for some night time fun in the woods.

I attended that day with high hopes which were soon dashed to the ground with amazing force. I practically ran to where the researcher had his booth set up and found him standing alone. He began to give me his spiel about the sasquatch and started by asking me if I had seen the Roger Patterson film which he had playing on his laptop. I told him that I had seen it before and then he began asking me if I had read any of the books that he had set out on the table. I advised him that I owned every book that he had there on the table and proceeded to introduce myself to him. I told him that I was also a researcher and handed him one of my cards.

I asked him if he was going to share any of the things he had found out in the woods during his presentation and he actually told me that he had never done any field work or had any interactions with the sasquatch. I couldn't believe what he was telling me. He said the only things he had to share with the crowd were other people's sighting reports and the things that he had read in the books that he had displayed. I really had a hard time wrapping my head around the fact that he called himself a "bigfoot researcher" but had spent absolutely no time in the woods with them at all.

As I stood there trying to figure out what to say next a woman in the crowd asked a question about how the Bigfoot gets along with dogs and this man told her that he had no idea since he had never read a sighting report that mentioned the subject. I proceeded to answer her question for her and told her that I had seen numerous private pictures showing dogs that had been killed or wounded severely by them and yet other dogs go out in the woods and come out completely unscathed. I told her I guessed it would depend on a lot of factors including the attitude of the dog and the attitude of the sasquatch at that time. I then found myself surrounded by people with questions that this man seemed to have no answers for at all. I answered what I could from my experiences and then left him to it. I was so disappointed that this "experienced bigfoot researcher" turned out to have even less knowledge than me and that I wasn't going to learn anything new from him.

In October of that year I got a call from an old friend named Cowboy who wanted to come down and do a weekend camp trip with me. I didn't want to go to the same area that I had been in all summer so I pulled up the map and searched for a new place for us to hang out. I decided on an area roughly fifty miles from my house alongside a river that looked like it would be perfect for a clan of sasquatch. It had running water, a large population of deer and elk, and a forest that went on for miles with no clear trails or roads running through it. I figured that even if we didn't have interactions, the weekend would be spent in some beautiful country.

We arrived on Friday evening with enough daylight left to set up camp and do a little hiking through the woods. This part of the forest was perfect for hiking with very little vegetation on the ground. We were having so much fun that we actually hiked a lot farther than we thought we had and didn't get back to the truck until a short while after sun down. We actually made the last part of the trek in the dark as neither one of us had brought along a flashlight. We drove back to camp and got a fire started and began to make dinner. He had found a piece of wood that looked like a head, and set it up in camp before we started the fire. As we were sitting there eating, we noticed that the boulder behind us looked like a skull also. We proclaimed this area to be called "Skull Rock", and it was so!

Figure 46 Skull Rock campsite

I was hearing slight shuffling in the bushes on the hill above camp but told myself it must just be the night creatures waking up and searching for food and dismissed it. We sat in our camp chairs in front of the fire after we ate and I kept hearing sounds from up on the hill. This short hill came down to where our table was set up and the sounds seemed to be getting closer. My friend leaned over towards me at one point and said he kept hearing noises. I told him that I did too and at that precise moment we heard a small branch crack on the ground in an area behind us. There was definitely more than one something encircling camp that night. I wasn't thinking about bear since we had found no sign of scat in the area so I told him it was probably the small nocturnal creatures coming out in search of food.

He leaned over and grabbed a piece of bread that was left over from dinner and tore it into little pieces and threw them out as far as he could and said, "Okay, I'm going to bed." With that he walked to his tent and I went to mine. I fell asleep really quickly and never heard anything else that night.

We spent the next day hiking up the river and sitting on the many huge boulders that are in the river and along the shoreline. We found deep swimming holes with azure water and small waterfalls were everywhere. While fording the river at one point, I looked down and noticed a small snake swimming towards me. I have a fear of snakes so I took off running. I looked back and noticed that the snake was following me! I kept running and rounded a bend in the river and found myself on a long sandbar covered in small rocks and pebbles. I forgot about the snake instantly and started rock hounding. My friend finally caught up with me and we spent quite some time just looking at rocks. My pockets were a lot heavier on the walk back to camp that day.

After dinner was served and the camp was cleaned up we found ourselves sitting in front of the campfire again. My friend walked over to the truck and pulled out a case of beer. At that point I became slightly irked. He knows I don't drink beer so I reckoned that case was for his consumption. I don't dislike drinking in camp; I just don't condone getting wasted when you're out in the wilderness miles away from everyone else. I'd like to think that if any emergency were ever to happen everyone would be sober enough to chip in and handle the problem. I knew that he wasn't going to quit drinking until he passed out, so I resigned myself to that fact and let him get his drink on while I stayed clear headed. Once again we were hearing things rustling through the vegetation and in the trees behind us. He seemed to be getting more and more agitated as the night wore on and I kept telling him that it was nothing more than raccoons or field mice. I took the flashlight away from him. The more he talked of a vicious bear out there, the more I started to believe him. I knew the odds were not high on that being what it was but he was starting to play with my head.

I can't tell you how relieved I was when he finally stated that he was going to bed and wobbled off to the tent. I continued to sit in front of the fire listening as the sounds grew more frequent and louder. I began to think that maybe whatever predator was out there, be it bear or cougar, was getting braver now that one human had been separated from the pack. I don't know why I let him get into my head like that. I knew that I needed to quit staring into the light of the fire so I could get my night eyes back so I stood up and walked into the middle of camp with the fire at my back. The fire was far enough away that I was no longer in its circle of light, but close enough that I could run for a burning branch if I felt that I needed it.

I had been holding my wand light while I was sitting by the fire and I carried it with me when I walked into the darkness. We had set up camp on the right side of a clearing that was covered in gravel and there was a small copse of Myrtle trees growing in the middle of this gravel area. As I was looking out across the clearing, I heard twigs cracking and gravel grating. The sound came from directly in front of me and ended in the middle of this tree circle. My mind immediately went back to bear, and I held up my wand and turned it on. It has a very powerful beam and lit the area up like it was daylight. I saw something jump up from the ground in the middle of two trees and run as fast as the wind off to my left side and enter the bushes there.

This happened so fast that the only visual I got was something hair covered running on four legs. My mind screamed BEAR! I followed it with the light as best I could but it ran faster than I could swivel my arm. I then heard the sound of something hitting, and then skidding across the top of the gravel at my feet. I was lowering the light down towards the ground to see what had just hit the ground in front of me when I saw a piece of gravel fly through the light beam in front of me and hit the ground and then skid across the top of the gravel.

I couldn't believe what I had just seen. The pebble didn't arc down from the sky like it was thrown overhand either. It came straight in sideways across the light from my wand. I knew then that it wasn't a bear that had come in, it was sasquatch! I couldn't describe how happy I was at that moment. I'd heard stories about people having things thrown at them in their campsites, but I had never had it happen to me! There was nothing scary about it at all. It was fantastic!

I instantly turned off the light and spoke to them out loud. I didn't even care if Cowboy heard me talking to myself in the darkness... besides, he was snoring so loud I don't think he could have heard me anyway. I remember saying, "Oh, I'm so sorry! I thought you were a bear! If I had known it was you, I would never have shined the light at you! I want good will and I mean you no harm. Thanks for coming in and saying hello." I then turned and skirted around our camp and walked over to where a small path led to the river. I heard walking sounds from three sides of camp and was feeling really good about standing in the dark alone. Then I heard a sound from across the river that started to make me a bit worried. It was the sound of something super big stepping into the river and walking across it. It only took three steps for this thing to be across the entire river and the fourth step was the sound of a large branch cracking on my side of the shoreline a short ways to my right hand side. I remember thinking that this thing must be huge, and I took a few steps backwards at that point. I then heard a very deep huff sound as though something was breathing out a very deep breath. It sounded nothing like a bear huffing, and I began to get a bit nervous.

I thought to myself that maybe this was the patriarch of the family coming to check up on what was going on. Whoever it was, they had to be much larger than the one I saw running away. I decided then that it would be most prudent to get myself back into camp. As I was turning to walk back into our camp, I heard this huge something walking through the trees to the right of our camp parallel to me. We both arrived at camp at the same time and I heard two very heavy footsteps walking up to the giant boulder on the right side of our camp and plant itself there. The only problem with that was that my chair was placed directly in front of that same boulder and I was planning on sitting in that chair. I had planned to sit down and try to look peaceful, and perhaps exhibit a showing of friendship and serenity.

I thought about how this huge somebody had walked directly to the boulder having to know full well that I was walking in that exact direction. I briefly wondered if it was a trap and if I should stop where I was and see what its next move was going to be... Should I continue on over and sit in my chair with my back facing something that could easily take me out, or even take me with it? Should I just change direction and stop in front of the fire and stand there like a dork? I quickly decided that whatever was going to happen would happen just as easily no matter where I was located in that campsite. I finally saw with my own two eyes how quickly this species can move, so what difference did it make if it had to take a few more steps to get to me?

I began to speak out loud again in a soothing voice and said how happy I was that the family had come down to meet me. For the large ones benefit I again repeated that I wanted good will and meant them no harm. I told them that I was going to come over and sit in my chair and enjoy having them around. I had reached my chair at that point and I will admit that I never looked up the entire time. I had been looking at the ground while I was walking and I continued looking at the ground as I sat down. My heart was hammering in my chest and I had to actively work at making it slow down. I could hear the large one breathing behind me and I kept listening for it to reach over the boulder to grab me. It stood there for approximately two or three minutes and I heard it walking up the hill behind me. At that point the woods erupted with the sound of snapping twigs and branches heading up the hill.

It got very quiet after that and I assumed that the family had followed this one.

I continued to sit in my chair for a long time with the biggest smile just plastered on my face. I had found another clan and they seemed to be okay with me being there. I felt as though I had been analyzed by the one behind the rock and found to be worthy. It was the best feeling ever and I was so glad that my friend had snored through the entire event. It made it seem more special somehow that it was just I and them that night. I knew I was going to come here as often as I could.

The next morning I scouted the camp site for tracks and sure enough there was a large area of matted down vegetation on the other side of that boulder. I also found a set of tracks behind the skull boulder. I destroyed them before my friend could catch sight of them. I then went to the shoreline and destroyed the tracks in the sand coming up out of the river. I never did tell him about our guest in camp and he never said a word about hearing anything in the night. We left a few hours later and went back home.

Chapter 18 – Back To Washington – A New Camp

I didn't make it back up to this area until the next summer. My girlfriend, Rhonda, came down to camp with me and we headed straight for the last spot I camped. Unfortunately, the spot was already taken by somebody else and the campsite looked like they intended to stay for a while. We continued up the road, checking for another cool place to set up our camp. I noticed that a lot of people liked to camp in this area in the summer and each place we checked out had someone next to it. Since neither one of us likes to camp with neighbors we had to drive quite far until we found a spot.

Figure 47 Our campsite beside the river

After checking more than a dozen campsites and finding them full, we pulled into the next campground to find it completely deserted. We couldn't believe our luck and had to go read the camp board to make sure the campground wasn't closed or something. There were no closure signs anywhere so we picked the biggest and best camp site at the end of the road and set up. While we were there numerous cars drove into the campsite and then drove straight back out again. We kept remarking about that. Were we so scary looking that no one wanted to camp next to us, or did they know something we didn't. We were just glad that none of them stayed and we had the whole place to ourselves.

The campsite was beautiful and sat right next to the river with a wide gravel bar to one side. There was a nice path leading from the back of the tent into the trees and along the river. We walked down to the gravel bar and headed for the river. We found a deep swimming hole with a waterfall coming down the mountain on the opposite bank. It was so peaceful and beautiful there. There are no pictures of this trip for me to share besides the picture of the campsite because as soon as I took the first two pictures the batteries in the camera died! I put my spare set in and they died also. I'd never had that happen before but couldn't say with certainty that both sets were fully charged to begin with so we'll leave it at that.

The clouds started blotting out the sun that afternoon and it started raining on us not long after dinner was over. We hastily began stowing everything away so it would stay dry. I opened up my large plastic tote and placed my wand light into it since I wouldn't need it any longer. We had other lights that we were going to be using while we were sitting in the tent.

I was sitting on the right side of the tent on my bedroll and Rhonda was sitting on her blow up mattress opposite me and we were talking and getting caught up with each other. The only sound we could hear was the rain beating down on the tent. She suddenly sat up straighter and turned her head to look behind her. I wasn't sure what was going on at that point until she looked at me and said that something had just poked her in the back through the tent. She shifted her position to be able to look at the back of the tent but I'm not sure if she ever saw anything. Nothing else happened and we eventually fell asleep.

The rains had stopped by the time we woke up the next morning and I about jumped out of the tent to heed the call of nature. I intended to walk around the backside of the tent and follow the trail into the trees for a bit before going, but as I rounded the tent I noticed something lying on the ground. It was the bright red which caught my eye and then I noticed that it was my wand light. It has a bright red rubber handle attached to it. It was laying on the ground right next to the tent on the side that I had been sleeping on. I knew for certain that I had placed that in my plastic tote because that is where it goes at night unless I'm using it in my tent. I'm a stickler for knowing where my stuff is because I despise always having to search for things.

About that time Rhonda came out of the tent and I picked up the wand and showed it to her and told her that I had just found it lying there on the ground. Neither one of us could figure out how it had ended up there. I turned it on just to be sure it was still working after having been in the rain and, fortunately, it's a lot tougher than I thought it was and wasn't damaged in any way. Then we started laughing about the fact that this could have been what was used to poke her last night.

The sasquatch had been known to open my totes and ice chests to see what was in them. The possibility isn't too far out there if you really think about it. They may even have been trying to poke me with it but were unable to, so they left it in the mud. When I sleep in my tent I placed my pack and other things between me and the side of the tent. It makes me feel safer and puts a slight barrier between me and whatever is out there messing with the tent. In other words, they could have been out there poking at my backpack while I continued snoring and remained completely oblivious to the entire event.

I called Cristy and told her about the events that had happened to us, and we instantly planned another trip to Washington. Both of us were missing the fun we had while we were in this area. This year we would be traveling not only with Connor but also with Cristy's new daughter Caydence and her ten year old step daughter. We knew it wouldn't be easy, but we were going to do it. We had to get back up to see the clan. It had already been a year and we were homesick for the place. It took us roughly nine hours to arrive

Figure 48 View outside my tent

at our destination after numerous stops for the kids and it was about dark by the time we pulled in.

There were other campers there when we arrived and they had taken both camp spots by camping in one and parking three vehicles in the other. This wasn't noticeable to us until we turned into the lake and tried to pull into our regular spot since it's hidden from the main road. Cristy stopped the truck and asked me what I wanted to do now, and did I want to try to find another place to camp. After the long tiring drive we had just made with three kids there was no way I was going to spend more time looking for another spot and besides, next to the isolated camp site that we could no longer get to, this was the only other spot worth having.

I told her to just put the truck in park and I got out and walked into the other camp site where everyone was sitting around the fire. I told them about the nine hour drive through hell that we had just made for the sake of camping in Washington and asked if they would mind moving their vehicles so that we could camp in the other spot. They all seemed completely agreeable except for one young woman with a chip on her shoulder. I just glared at her and told her that now was not the time to mess with me, not after the drive I had just made, and she quieted down. They moved the vehicles for us and we got the kids out and fed them dinner so we could get the tents set up. Everyone was really tired and cranky that night and we all went to bed early.

I wasn't holding out much hope of having the sasquatch show up around camp with the other campers next to us, but it was great to be at the lake again anyway. We only had to share the campground with the other campers for two days since we arrived on Friday and they left on Sunday. As soon as they left we shifted our camp over to the larger spot. We had three tents with us on this trip and needed the extra room so we weren't all sleeping on top of each other. I never realized how tiring it is camping with children. There is virtually no down time as they are constantly running around and trying to sneak away from camp when your back is turned during the day, and then always trying to get back up to sit around the fire with you when you put them to bed.

We also had the ten year old with us who liked to hide outside of camp and eavesdrop on our every conversation. We didn't want her to hear us talking about the sasquatch because we didn't want her to get afraid and refuse to sleep in her own tent. We also didn't want to ruin her vacation and make her spend the entire time feeling fearful when the sun went down. But we never heard a thing at night and nothing was changed in camp when we woke up in the morning. After spending a week there with no interaction from the sasquatch I started to feel a bit down.

I talked to Cristy about it one evening at the campfire and she mentioned that perhaps they just weren't up on the mountain at that time, or perhaps they were staying away because of the step daughter. Either they stayed away because they didn't trust her, or because they didn't want to scare her. I told her that it was beginning to get me down that we had driven all that way to spend time with them and they weren't around and she said that she felt the same way.

When I went to bed that night it was with a heavy heart. I was having a great time hiking around the lake and playing Ploop with Connor on the shore every day, but I was missing the action that the sasquatch would have brought every night.

When we had set up the camp this time I put my tent behind some bushes and Cristy put her tent in front of them so we had a bush buffer between the two tents. The ten year old's tent was approximately twelve foot away and to one side of mine. About one am that same night I was laying on my right side on top of my bedroll and facing the wall of the tent to my right. I could hear the sounds of deep breathing coming from the other two tents so I figured everyone was asleep already, but I was still wide awake. I then heard the sound of a hand being dragged across the opposite side of the tent, although I hadn't heard the sounds of anything coming up to it.

It sounded like fingers sliding across the nylon from front to back and then the sound started sliding across the back of the tent! As the sound reached the back of my tent where the window is located, something gently pushed on my left shoulder and actually rocked me forward slightly. I actually felt a hand push on my skin since I was wearing a tank top at the time and nothing was covering my shoulders or arms.

I immediately rolled over on my back and looked at the back of the tent. I didn't see any change in the material and I didn't see anything coming through it. It all happened within a few seconds and I just couldn't figure out how something outside of my tent could have possibly touched me inside of my tent. I briefly thought that perhaps I had left the window open and they had reached in that way but as soon as I looked at the zipped up window I knew that wasn't the answer. Besides, I have a screen over the window so even if it was unzipped, they still couldn't reach through it. I then heard the sound of something rustle the bushes behind me and then walk across the gravel and off into the night.

I kept wracking my brain trying to figure out how something could have touched me inside my zipped up tent and I started thinking that it must have been something in the tent that brushed against my shoulder. That's not what it had felt like, but what else could it have been? The more I looked, the more I realized that it couldn't have been something already in the tent because there was nothing there to explain that theory. Everything was piled up against the sides and the middle was empty. There is nothing that hangs down in the tent either.

I then thought that maybe they had pushed the tent in far enough to touch me that way, so I started to experiment. I pushed on the back of the tent to see how far out I could push it. I soon realized that this wasn't the explanation either. The entire tent would start to buckle on the sides if I pushed it out far enough to explain what I was thinking and that didn't happen when I was touched! I couldn't figure out how something solid could just reach inside my tent and touch me with its hand. I was so perplexed about how this could have happened that it took me awhile to fully realize that it had actually happened!

When I finally quit obsessing over how someone had managed to do this, I thought about how I hadn't heard anything walk up to the tent, but I sure had heard them walk off afterwards. I decided to get out of the tent and take a look around. I wanted to see if the little girl had awakened and came over to mess with me, but she was still fast asleep. I then checked to see if there was enough room between my tent and the bushes for a sasquatch to get behind it. It turned out that there was enough room and that it could have been possible. It did look as though something had walked on the ground vegetation behind my tent, but I couldn't be certain that it wasn't caused by me setting up the tent in the first place. Of course that would have been a week prior to this night. There was no dirt or other markings on the outside of the tent that I could see either. I looked in the direction that I heard it walking off in and gave it a gesture of respect and said, "Thank you for coming to visit." I went back into my tent and laid there for quite some time hoping it would come back but nothing did and I eventually fell asleep.

I told Cristy about it the next day and she said that she did hear something rustling the bushes between our tents and had hoped it was the sasquatch. I can't think of a better explanation for this happening since they do seem to be capable of doing the wildest things and I'm not prepared to believe that ghosts roam the forests of this area. I don't know what the touch was for, but I'd like to think that they knew that I was missing them and came in to say hello, but who could be sure about why they do what they do. I just wish I hadn't panicked at the time and maybe he/she would have stayed longer.

We stayed for another four days but never had any other contact that trip. I haven't had the opportunity to get back to Washington again and I can only hope that they are all still safe. I can say that I will always look back on the time I spent there with a heart full of love, and I will always miss them. They always treated me with respect and love and I will be forever grateful.

We arrived home from this trip a bit more sorrowful, and felt like we were leaving behind our best friends. Cristy called me a few days after we got home and told me about how Connor had woke up from his nap that day and she had gone in to get him up. When she did, she noticed something very strange on his blanket. It was a wet spot on the blanket that just so happened to be in the shape of a sasquatch. She said that she just could not believe what she was seeing and she took some photo's to send to me.

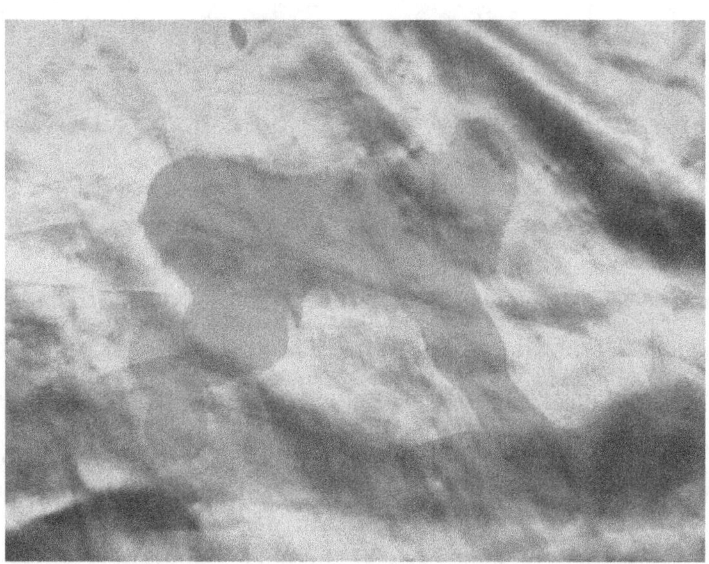

Figure 49 Stain on Connor's blanket

When the pictures arrived in my inbox, I was completely shocked to see that it did indeed look like a sasquatch which was down on its hands and knees! The stain was complete with a beard, very stocky arms, and what appears to be hair hanging off its middle and hands and feet. You can even see the muscles on its arms. Also, the arms are very long and the legs are shorter.

I called her back and told her how amazed I was at this picture, and she said that she was amazed when she found it. She has never washed this blanket and the stain is still on it to this day.

It has never faded or bled at all. It is still crisp and distinct and I'm still amazed by it.

Chapter 19 – Skull Rock

While attending the Beachfoot gathering, an outdoor conference and symposium on the subject of sasquatch, that summer I was introduced to the fact that some of the people I knew were studying glyphs as they pertained to the sasquatch. Brian Bland and Thom Cantrall gave a presentation showing a very high number of stick signs they had been finding on the ground up in British Columbia. I had never noticed sticks lying on the ground in any sort of pattern before and seriously had my doubts about the fact that sasquatch had put these together, but they showed pictures proving that they had found the same style of stick patterns in various areas including here in the United States.

We all know that sticks do fall from trees and down onto the ground. Invariably some of those sticks will fall on top of each other and form patterns. When these sticks are woven together, on top of and under each other you have to look a bit closer. This had never been something I'd looked for in the woods and I had to ask myself why I hadn't been. Hearing this presentation reminded me of how I had found the white carnations on the headstones in the cemetery that were formed into that AH pattern when I first began looking for the sasquatch in 2008. This fact had slipped my mind entirely after all the stranger things began happening to me. I listened carefully to the presentation and kept an open mind to the subject.

On the last day of this get together we all began breaking down our camps and loading up our vehicles while still wandering around and saying goodbye to our fellow researcher friends. I wandered over to Thom and Brian's campsite to see if I could be of any assistance to them. We were standing in a huddle talking when one of the people in this group stepped over to the side of her tent to release a stake from the ground. I heard her gasp and turned to look at her and saw that she was looking down onto the ground. I let my eyes fall to where she was looking and I saw that there was a stick sign laying on the ground underneath the rope where it had been staked into the ground.

Figure 50 Glyph left under tent line

I bent down to look at it as she was exclaiming about it to the rest of the group. It was created with small sticks that were indeed woven under and over each other. At that point the place was filling up quickly with other people who wanted to take a look at it so I backed off and gave them all room to do so. I couldn't believe that a sasquatch could have made this, but I totally trust and respect the persons who were camped in this area and don't believe that they would make something like this up. I will admit here that I still had some doubts about it and went off to think this thing through.

I had heard rustling in the brush behind my tent the previous night followed by a deep exhalation of air before something walked off, so it was conceivable that they had been in the campground. It was surrounded on three sides by the river and the fourth side led straight up into the mountains so they had great access to the place. It was occupied by people who knew of the sasquatch and some who even had habituations taking place at their residences. So why couldn't they come in to say hello to a few people?

I kept thinking about the possibility of the glyphs being a form of communication for the sasquatch while on the long drive home. I finally told myself to log the information into my mind and let it be. If it was fact, I would run into it one day and if it wasn't, at least these people were spending time in the great outdoors and having a lot of fun doing so. I couldn't say whether it was fact or not simply because I hadn't had the experience yet. I think there are far too many people out there screaming 'hoax' simply because they just haven't experienced it for themselves.

Cut to a few months later when I went camping at Skull Rock with Cristy and Rhonda who gotten poked in the back in the rain. Since we had three tents with us this time we had to set up the camp a bit differently and my tent ended up being erected right next to what became our trail from the truck to the picnic table.

Every time we unloaded something from the truck or walked out to the porta-potty, we used this trail. Obviously, this trail was used extensively and well into the night.

Cristy placed her tent a short ways out of camp under two trees right at the top of the trail that leads down to the river where I stood the night I heard the big one coming across. Rhonda put her tent up in the middle of the camp where mine had been erected on prior trips here.

The first night we were there we could hear the crickets chirping all around us then they would all become quiet. We heard branch snaps around the outside of camp and we could feel a presence out there watching us. We eventually dowsed the fire since it was fire season and didn't want to leave it unattended and crawled into our separate tents. As I was sitting there reading, I heard something walk up to my tent.

I was so excited but I didn't say anything aloud and waited to see what it would do next, or if others would walk into camp at that time. I then heard it walk away from my tent in the direction of Rhonda's tent and smiled knowing that it was making the rounds around camp to say hello. I sat up listening to the sounds of leaves rustling and gravel popping around camp and then it got really quiet. The crickets still hadn't started up again so I figured that the sasquatch must still be around, but we had had a very long day and I couldn't fight sleep any longer.

The next morning I fell out of my tent to find Rhonda already up and drinking coffee. Cristy and the kids hadn't made a sound yet, so we walked down to the river to talk and allowed them to sleep a little longer. I didn't make the trek all the way to the outhouse that morning and never used the trail that ran beside my tent. While we were down on the river I heard Cristy call my name really loud. When I answered her she told me to come up to her tent so she could show me something.

When Rhonda and I arrived at Cristy's tent, she was looking at the ground. I looked down and saw a glyph! It was located very close to her tent at the top of that trail. She said that it wasn't there when she got into the tent the night before and she never got out of her tent during the night at all.

Figure 51 Glyph in front of Cristy's tent

None of us could really believe what we were seeing... I less than any of them. This pattern very closely resembled the AH pattern that I had seen in the cemetery! The difference is that the top of the A here didn't meet up like the others had, but since they were formed with two flower stems, perhaps the A had to meet up in order to continue the pattern and they didn't have that problem using the sticks. Who's to say really, since I don't have a clue what it means.

How much of a coincidence is it that we had both been wrestling with the concept of them speaking through pictures left on the ground and now we find one right in front of Cristy's tent. I ran to get my camera from my tent and stopped dead in my tracks when I reached it... I had approached my tent from the back and was intending to use this narrow trail to get to the front of it. As I reached the back of my tent, I noticed something very strange laying there on the ground.

There was a stick pattern lying there also... And not just any ordinary stick pattern. This one had been formed using beach grass along with the sticks! It was completely different from the one left in front of Cristy's tent, and I remember getting tears in my eyes when I found it. No one else in the world knew what this beach grass meant to me... especially in conjunction with the sasquatch! Only they could have known what I would be feeling when I set eyes upon it.

The beach grass was woven up and under the sticks that were also laid there, so it wasn't just dropped to the ground. It had to mean something, but what? As can be seen by the pictures of both glyphs, there was no beach grass

Figure 52 Glyph left by my tent with tracks

growing in this area. As a matter of fact we were some seventy miles or more from any beach and we hadn't found any swamps in this area at all.

It was left next to my tent directly on the trail that we had been creating by walking through here the day and night before. That means that it couldn't have already been there when we arrived or we would have kicked it all over the place and scattered the grass. It wouldn't still be lying there all woven together like that at all.

Once again, I had been shown something that I was full of doubt over. I was given the experience so that I could believe it and once again they had done this by throwing it into my face. I keep saying it's pretty hard to doubt something you've seen with your own eyes but it keeps happening to me. This time they had proven to me not only that my friends were telling the truth, but that this was indeed something that the sasquatch people did. Now if I could only find out what it means and what the glyphs stand for.

Are they trying to tell us something or is it just their calling cards identifying each different clan member? If it is just a calling card identifying them, why did they use the beach grass on mine and not on Cristy's? It couldn't have been the same clan that I had run into in the sand dunes of Washington, so how did they know to bring beach grass? And where did the grass come from? Surely not

Figure 53 Thom and Kathi at Beachfoot

from the banks of the river we were camped on. Everyone agreed with that point as there simply wasn't any growing anywhere around us.

I took three pieces of this sharp beach grass and put it into my tent to take home with me. I knew this sign was going to be trampled by us, or the kids, at some point since we walked this path often and I wanted to keep some for my memories. It is far too easy for my mind to start forgetting the little things in light of the bigger things that keep happening... Recall how I had almost forgotten about the flowers in the cemetery. As I well know, pictures can be lost!

For some reason we had no other action that weekend. I was left to wonder if this was their way of saying goodbye or something. I think we were all a bit disappointed that they never came back. This campsite was full for the rest of the summer and we had to camp up at the site of the back poking episode from then on, but we never did have any other activity. Seems that fall is the best time to be up in this area.

Chapter 20 – Back to Skull Rock

I absolutely love when I can make a believer out of someone else by giving them their own experiences. This is exactly what happened to one of my new friends, Ladawna I met in Coos Bay. She had been listening to me go on about my experiences and looking at the many sasquatch pictures and collectibles that I have around my home until she expressed an interest in going camping with me.

We made plans to head into the mountains on a Monday evening in October. I knew we would have the place to ourselves and the mountains are so beautiful in the fall that I love to go camping then. We set up camp in the site that I have dubbed Skull Rock. Of course, we had to clean the campsite up before we set our stuff up and bury a deer carcass with leaves and dirt that was left laying off to one side. Please…don't pollute our forests!

It was a beautifully sunny, fall day when we arrived. After we set up camp we walked along the shore until we found a nice spot to sit and look across the river. We were just talking and enjoying the fall scenery. I hadn't told Ladawna much about anything that pertained to the sasquatch... just the basic facts that anyone can learn by watching the various TV shows on the subject. I was very surprised when she began speaking in a hushed voice next to me. In fact, it sounded like she was talking to herself more than she was talking to me.

I heard her say, "The forest just moved…no IT didn't move! Something in the forest must have moved. I was just looking out the corner of my eye and the forest shimmered and moved. But it looked like the trees were moving."

I turned to look at her and then looked where her gaze was falling but I didn't see anything out of the ordinary. I asked her what she just said and what she just saw. She said it looked like a heat shimmer and she didn't see it when she looked right at it, only from the corner of her eye.

I found that to be a very surprising thing for her to say. Especially since I know that she doesn't follow any bigfoot forums and I know she hasn't read any of the many books I have on the subject so how would she even know about the numerous reports of this type that have been published. I believe that she saw something invisible moving through the forest, like the movie Predator. I know that there are hundreds of reports where others have reported this type of anomaly in association with the sasquatch... I just haven't seen it for myself yet.

We discussed this for a few minutes and then made our way back to camp. We had no further interactions with them that night. The next day it began to rain, so we set up the tarp over the top of the picnic table and played a few hands of cards until it was time to make dinner. This is the same picnic table that sits at the bottom of the slight hill that runs down into camp. I talked about it in the previous chapter about the Skull Rock camp site.

The sun had already set and it was a very dark night. I was standing at the end of the picnic table slightly under the tarp, cleaning up our dinner and putting things away into the cooler which we had tucked under this end of the table. My friend was only about ten steps away from me and was standing next to the fire pit when we suddenly heard a high pitched scream coming from the hill directly above me!

Figure 54 Our camp... note the small hill to left

It was followed very closely by a second scream from the same individual. They were two very quick sounds. I had never heard this type of scream before. It didn't go on like a siren, like the vocals I had heard in the Bluff Creek area. These were quick, short bursts of sound that were very high pitched and made me think it must be a female on the hill above us.

I instantly looked at Ladawna and she was looking at me. I tried looking up on the hill but it was too dark to see anything there. This hill wasn't silhouetted by the sky because there is a very high, rocky wall across the road which blocks out the sky completely. I could see nothing but inky blackness up there. I asked her if she had seen what it was and she said that she hadn't, but she had certainly heard it.

It was then that we heard the vegetation on this hill rustling a bit to the right of our camp. I heard gravel clicking together as something crossed the gravel circle on its way to the river. Then all was quiet... I thought maybe it was hungry and had screamed for food so I quickly threw some of our dinner on a plate. I began walking to the truck while speaking out loud and saying that we wanted good will and meant them no harm and that I was going to share some of our dinner with her if she wanted it. I placed the plate on the hood of the car and then walked back into camp.

My friend told me that she had heard the wailing sound coming from farther up the river and that she thought this one had wandered away from camp. I didn't think so, but never let on. I had a feeling that this one was going to be joined by others, since that's the way it usually worked. I had already set out white rocks around the tent to signify friendship, but I hadn't told her what the meaning was. As I've mentioned, I'm pretty secretive with people. I would rather they have their own encounters and not put thoughts into their minds and give them ideas about what could possibly happen.

We got into the tent early that night because of the cold and the rain. I was hoping she would lay awake in her bag just in case they did come into camp, but I heard her softly snoring moments after we lay down. As soon as I heard her start to snore quietly, I heard footsteps coming into camp behind the tent and walk to the picnic table. I was hoping the sasquatch had come back and eaten dinner, since she had to have passed the truck to get into camp from that direction.

I then heard the ice chest move and immediately I heard something vocalize in front of the tent. This one was standing right at my feet outside the tent door! I got a smile on my face knowing that there was more than one in camp at the time. I thought about waking my friend, but to do so I would have to turn over in my bag and then lean about four feet over to tap her, and I didn't want to make any noise that would scare them out of camp.

I was just lying there listening to the sounds of one of them moving about camp when they began to speak to each other. I had never heard them vocalizing in this manner before, and I had never heard it being discussed anywhere. I began to wonder if it was the sasquatch in our camp that night, or an Extraterrestial of some sort!

They kept vocalizing back and forth to each other and they weren't being quiet about it at all. I kept hoping they would wake my friend up so she could hear it and truthfully, I don't see how they didn't. I had one standing outside the tent at my feet and the other one was right next to the side of the tent that she was sleeping on. But she never did wake up.

I know they were in camp for over ten minutes and it was the wildest conversation I had heard to date. They sounded like they were just standing there shooting the breeze, like two women who meet at the back fence and have nothing better to do. They didn't sound hurried, or even concerned in any way that we might be listening to them. It wouldn't surprise me in the least to find that one was casually sitting at the picnic table, while the other was leaned up against the skull boulder just as calm as can be discussing the day's events!

If you are wondering why I'm not accurately detailing the type of vocalizations they were using that night, it's because I have never heard these types of vocals ever described in any forum, or in any book that I have ever read. I don't want to give the hoaxers any more information than they have already. I don't want to feed them so that they can lie about their interactions with this amazing species of life, or give those who call blast in the forest another way to try to lure them in.

I'm sorry if this offends anyone, but this is as it must be to protect my friends who live in the forest... And I surely do consider them friends of mine. I have never been threatened by them in any way, and I will never willingly do something to threaten them or their families in any way.

The next morning, I told my friend everything that she had missed the night before and I'm not sure if she really believed me at that point. I think she wanted to, but I admit that some of it may have sounded a bit far-fetched to her. I can't blame her for this, I had a lot of problems believing some of it myself, and I was listening to it.

We went for a walk around the area looking for evidence. Since it was autumn and most of the ground was covered in fallen myrtle wood leaves, it was really hard to find anything like prints or trackways. Ladawna did find a large print that was left in the gravel circle and took a picture of it. She expressed interest in coming back to this location again, but between working and raising three kids, she's been too busy make it. I do know that she is now a believer!

I'm glad that I had the opportunity to open the eyes of so many people around me to the reality of sasquatch. They truly are a remarkable species. And they truly can do some remarkable things.

Figure 55 Footprint in gravel

Chapter 21 – Bluff Creek

Everyone who's ever taken an interest in sasquatch knows about the Bluff Creek area in California. It is the area where Jerry Crew and his workers found and reported hundreds of sasquatch tracks on the road near, and around, their machinery. They also had some very heavy oil drums and excavator tires thrown off the roadway. When this was reported, the press dubbed this species "bigfoot" because of the size of its tracks. It is also the place where Roger Patterson and Bob Gimlin recorded the female sasquatch walking along a gravel bar. Although this footage has been analyzed by many different people, and has created many arguments within the bigfoot research community, people are still divided over whether it is authentic or a hoax. I am here to give you my assessment of the film. I think it's real and I think that anyone who doesn't think it's real has never seen a sasquatch in the flesh... But that's just my humble opinion.

I had wanted to visit this area since I first became acquainted with this species. For me this was the mecca for sasquatch research. It was ground zero in my mind and I could only hope and dream of ever having the opportunity to camp here. My prayers were answered when Thom Cantrall asked me if I was coming along on his camp trip that summer. They were going to be camping on Blue Creek which is very close to Bluff Creek. I just could not believe my ears when he said that and I allowed myself a moment of elation. However, that moment soon ended when I realized that I was in no financial state to make a journey like that. I had to decline on the grounds that I couldn't afford it.

Much to my surprise he said that he wouldn't hear of me staying home and arrangements could be made which would allow me to tag along. I do believe I floated to the ceiling at that point and whooped for joy. I was going to be able to spend some time in the Bluff Creek area and we would also be driving into Willow Creek to visit the Museum there! It suddenly felt like Christmas again, and I was on cloud nine!

It was decided that I would meet everyone in Roseburg, Oregon and I would be driving down into California with Sue, a friend from Eugene, OR, in her car. We got a late start that day because we were waiting for Thom to arrive from the Portland airport where he was picking up Jackie, a very nice lady who flew in from England to join us on the campout. Since it was getting close to dark already, it was decided that we would find a spot to

Figure 56 Me with statue in Happy Camp

camp for the night and get an early start the next morning.

We followed Thom to a wide gravel circle in an area he found and we all set up our tents for the night. I spent some time trying to get a small hobo fire started and at one point someone snapped a picture of Thom sitting in his tent. He was surrounded by orbs! In the picture you can see that the moon was just riding on the crest of the mountain and it wasn't raining.

I completely enjoyed the long drive down. The company was great and the scenery was absolutely beautiful. We stopped a few times along the way to take pictures. When we made it to Happy Camp, Ca. we stopped at the statue there to get out and stretch and take more pictures. This is a huge metal statue of a sasquatch and I wanted to hold its hand, but I could barely reach it.

After we left Happy Camp it was on to the G-O road! This is the road that Jerry Crew and his workers were building when they found the tracks. It is actually the Gasquet-Orleans road as it was intended to run from the town of Orleans, CA to Gasquet, CA, but it better known as the G-O road. As soon as I saw that sign I began giggling like a school girl and had to take a picture of it. I still couldn't believe that I was even here, and I remember feeling happier than I ever had before.

The countryside is so wild and rugged, and yet so peaceful and beautiful, that I started to get an ache in my neck from whipping it around so much and trying to take in all the beautiful scenery. We drove up on a couple of two year old bears crossing the road in front of us and we all slowed down knowing that mama could be coming up behind them. We surely didn't want to hit her with our cars.

We drove along the G-O road for a while talking about how this was the perfect habitat for the sasquatch, and how strikingly gorgeous it was out there. We finally turned down what would be the last road before our designated camp spot and I just couldn't believe that the scenery got even more beautiful than out on the main road. This truly is a gem of a wilderness area.

We followed a very narrow mountain road that fell into a giant meadow. I remember gasping when I first spotted it because of its beauty. I was secretly hoping that this was where we were going to be staying, and I was thrilled to see Thom pull off to the side of the road and proclaimed this as our new home for the next twelve days. Everything was so green, and the sun was so warm, and there were so many flowers growing in the grasses that I just stood outside the car for a few minutes trying to emblazon it into my memory for future use. It's exactly what I would picture when someone said, "Go to your happy place," and now I was actually standing there and seeing it in person!

Everyone started walking around and designating different camp areas and deciding where to put their tents. I just hung back because I knew in my heart that wherever they set up, I was going to pick an area far away. Not because I didn't like anyone there, but because I wanted to be far enough away to have my own experiences with the sasquatch that lived there. I knew in my heart that we were going to be having some wonderful experiences out there.

Judging by the people who were going to be camping with us in this meadow, there was no way we wouldn't have a few come in and say hello. I wanted to be far enough away from everyone else that if the sasquatch did come to my tent it would be because they wanted to come to my tent and not just on their way to someone else's tent. I eventually set up camp quite a distance away and on the other side of the meadow in front of a tall tree.

Figure 57 My little tent placed across the meadow

We had so much fun greeting everyone as they arrived and I couldn't believe how wonderful and kind everyone was. You could just feel the love and happiness in the air. I know this makes it sound like some kind of hippy commune, but it wasn't that sort of energy at all. These were just genuinely happy people who loved unconditionally, and it showed and I felt it. At first I felt a bit out of place because I had been feeling a bit blue about my life recently and I wasn't emitting the same kind of energy as they were, but they all treated me like one of the family and I will be forever grateful for being invited along.

The next morning a group was talking about taking a hike up the mountain. I was thinking of joining them when Thom asked me if I would mind riding into town with him to pick up some supplies. Since I was eager to see some of the surrounding towns, I gladly accepted his invitation. We drove into Orleans and checked the store there, but as Thom remarked, "We'd have to visit the bank and take out a loan just to buy ice here!" So we drove on a bit farther up the road before finding a store at which to buy our supplies. We didn't arrive back at camp until around five pm and we sat down to dinner with the rest of the campers.

One of the ladies in our group talked about having seen a light colored sasquatch earlier in the day while they were hiking. Since we hadn't heard anything around camp the night before, I was hoping that it had followed them down the mountain and noticed us camping here and then went off to tell the rest of the clan that we were here.

Since some of our group had come in from Oklahoma and had a different time schedule, they retired early each evening. The rest went into their tents around eleven pm, but I am used to staying up late in the forest and I found myself alone around the fire each night. I would just enjoy the night until I was too tired to do so and I would put the fire out and wander off to my tent. There were strict fire warnings in the area at that time so the fire had to be put out completely when unattended.

That night as I stood alone by the fire I heard a rustling in the grass and looked over to see a small field mouse running through our kitchen area looking for scraps. He was such a little cutie and I felt bad for not having anything to drop for his dinner.

All the food was put away from the bears and I didn't want to pull anything out. I put the fire out and walked over to my tent and hoped he had found a little something for himself.

I hung my wand up at the top of the tent and then lit two small votive candles for a bit of warmth. I wanted to read for few minutes before crawling into my mummy bag and falling asleep. I was sitting there reading with my back to the back of the tent when I began to hear a strange buzzing sound inside of the tent. Yes, I said inside of the tent!

It started behind my head and then moved off to my left and to the front of the tent where it swirled around and then came back on the same path to the back of the tent behind my head. It was the strangest thing one could hear in the woods because this was around one am and it sounded like a cross between a bumble bee and a hummingbird. Not quite a buzzing sound, and not quite like wings beating at the speed of light, but a perfect mix of the two.

When I first began hearing it I thought maybe a bee had gotten into the tent somehow and I was about to grab for the light off the ceiling of the tent and thought maybe I could shoo it outside. Then the sound changed and sounded more like a hummingbird's wings beating rapidly. I was even more confused. Beside the fact that I never saw a hummingbird flying at night, it would have been large enough for me to see it flying around the inside of my tent and there was absolutely nothing there for me to see.

After the sound ended I just sat there for a while wondering what the heck that could have been. I then began to feel the bitter cold that had settled down on the meadow that night. I was just beginning to slide into my mummy bag when I heard the sound of branches snapping in the tree line to the right of me. I instantly froze so I could hear it if it happened again. A few minutes passed and I heard nothing, so I again started to slide into my bag. I then heard another branch snap and it sounded closer this time. I froze again and it got quiet again.

This happened a half dozen times until I became so cold that it was either I just slide into my bag and take my chances as a sasquatch burrito, or I just sit there and become a prime candidate for hypothermia. As soon as I would move, they would move. As soon as I would stop, they would stop. To me this felt very sneaky and I thought maybe they were up to no good out there. This went on for ten minutes or so, until I just decided that it didn't matter what was happening outside my tent, I had to get into my bag and warm up.

I then lay there listening to the sounds of two different sets of footsteps walking out of the woods on the right side of my tent. I heard the sound of rocks grating together with each step, which led me to believe that they were walking down the dry creek bed located next to my tent. I assumed they were heading into the area of our camp kitchen, and the other tents, so I rolled over and got comfortable and fell asleep.

When I awoke the next morning, I asked everyone sitting around the fire if bees or hummingbirds ever flew at night. They all looked at me like I was nuts and pretty much dismissed the question. I told them that it had sounded like something was buzzing around in my tent at around one am in the morning. I don't really remember anyone offering many suggestions as to what it could have been.

I joined some of the ladies that were hiking up the dry creek bed that came down the mountain and ran into camp on the right side of my tent, and we found a nice little watering hole about half way up the mountain. Since it was a very warm day, we all sat and enjoyed the coolness of the spot and took lots of pictures. One of the ladies with us said that she had just spotted a watcher, and we all said hello and then hiked on back down the mountain into camp. It was so nice to sit around the campfire with everyone that night and listen to the drums being played along with some very nice voices ringing out into the forest.

I'm not going to recite the day to day details as they happened in camp, simply because it would take me a couple different chapters and most of it was spent hiking or sitting around the fire talking. Suffice it to say that we all had a marvelous time laughing with each other and listening to those with musical talents entertain us. Actually the time ended all too soon. I could gladly live in this meadow in the mountains, and probably not miss interacting with people all that much. Every area that I walked to was more beautiful than the last, and I didn't believe that could be true when we first arrived and I set eyes on the meadow.

Anyone who knows me at all knows that my favorite day up there was the day that everyone elected to go into town and I got to stay up on the mountain by myself. As I stated before, it wasn't because I didn't like everyone else, it was because I'm basically a loner at heart and I feel so free knowing that I am the only one around for miles. This way there is no one around to think I'm doing something wrong or foolish or to berate me for it.

I grabbed a cup of coffee and hiked up the mountain. I eventually found a huge boulder wedged in between two trees. It looked so magical sitting there by itself that I just sat down near it and finished my coffee... Seemed like the thing to do at the time.

Figure 58 Boulder on mountain

On my way down the mountain I got that all too familiar feeling of being watched... the feeling that stands the hairs up on the back of my neck and makes me feel uncomfortable. I thought about all the bears we had seen... not only on the way into camp either. One of the ladies we were camping with had taken a walk down the road which led out of camp a few nights before my lone hike and she had come back in a big hurry. She said that she had walked down the road and had spoken out loud asking the sasquatch to show themselves to her. Then she heard a rustling in the brush alongside the road. Much to her dismay it wasn't a sasquatch that ambled out onto the roadway, it was a bear. She hightailed it back to camp and it obviously continued on its way. This was only a few short yards from camp.

So I thought about this as I walked over to a big boulder and climbed up on it and began to sing loudly.

I had always been told to talk when walking through the forest to alert bears to my presence. That way I wouldn't just stumble upon one and scare it into attacking me. I figured that the sound of my voice would alert whatever it was out there that there was indeed a human in the area, and the sound of my singing voice should have hurt the ears of anything listening.

I finished my short song and listened to the sound of absolutely nothing. No bird song, no rustles through the trees... nothing. So I spoke loudly and said, "I just sang a song for you, now you sing a song for me." No sound came out of the forest. Of course, I'm not sure what I would have done if something had started singing back to me, so maybe it was for the best.

I walked off the mountain and then along the creek that ran behind our kitchen area where I found another large meadow surrounded by trees. There were a couple different watering holes along this creek and I delighted in sitting next to each one just watching the butterflies and various other insects that would come in for a drink. Each one had its own unique beauty, and I took dozens of pictures to remember them by. I even found the source of the creek, where the water came up out of the ground. This area was so full of life that I couldn't resist the temptation to just sit in the sun there for an hour or so and enjoy myself. I took a couple videos at this spot also, and the bird song was simply amazing.

I then ambled over to lay in the hammock one of our campers was sweet enough to erect for us. It was located inside the tree line to the right of my tent and as I lay there just soaking up the warm California sun, I again got the feeling of being watched. Well... more like scrutinized for some reason. This time I just laughed and told them, "I'm on to you. Just come in and say hello." Once again I got nothing in return and I just continued laying there and singing whatever songs came to mind.

I had been having the time of my life and I never wanted the day to end, but of course the sun started sinking below the tree line as it is wont to do each day. I walked over to the kitchen area and got a fire started and waited for the others to get back.

One evening shortly after we arrived at this meadow, I was standing at the fire alone when I heard a slight rustling of the grass to my left. I didn't want to whip my head around in that direction and alert whatever it was to my presence. I hadn't heard the unmistakable sound of a bear approaching, so I figured that I was in no danger at the time. I was looking towards the fire, but keeping an eye on that side of me, when I saw a large deer wander into camp. It walked right up to me and stood next to me on my left side. It just stood there looking into the campfire with me.

This surprised me to no end. I turned my head to look at it, expecting it to bolt back off into the trees, but it just looked up at me and then back down again. I wanted to reach out and pet it, but I didn't want to make any sudden movements and scare it off. I just stood there with a huge smile on my face and looked at it. It was a beautiful animal and I just couldn't get over how it just stood there next to me like that. And not far away from me either. If I had just slightly leaned my body to the left, we would have been touching.

It was a remarkable experience and I didn't want to do anything that would make it end prematurely. After standing still next to me for about eight to ten minutes, the deer just looked up at me and then turned and walked slowly out of camp in the same direction that it had entered. I lost track of it after it walked out of the fire light but I could hear it rustling back through the grass again. I'd like to think that he looked up at me and said goodbye and then continued on with his life. I will never forget that time either, it was just so wonderful.

That night I got into my tent and was reading with the light on and I heard the sound of footsteps approach my tent. I quit breathing and sat still to see what they would do. They had to know that I was still up because my tent was lit up like the mother ship at the time. The footsteps stopped a few feet from my tent and then I started to hear the sound of vegetation being plucked. Not ripped up out of the ground, but plucked. Like when one sits on the ground and plucks out a blade of grass one at a time.

This really confused me. For a brief time, my mind tried to tell me that it was the deer and that he/she had come back to eat some grass. But I knew beyond a shadow of a doubt that I had heard the sound of heavy footsteps and nothing else in the world sounds like that. Especially when they are so heavy you can almost feel them on the ground. So I laid there trying to figure out exactly what was going on beside my tent... Though I will admit that I never once contemplated unzipping the tent and poking my head out for confirmation of what I was hearing.

This went on for quite some time and I started getting tired and wanted to climb into my bag, but once again, I didn't want to make a wrong move in case this someone had thought I was asleep. I'm pretty sure it's bad form to surprise or irritate a species this large! The longer I sat and listened, the more sure I became that it was the sound of vegetation being plucked. Then, whatever was out there walked back into the tree line and I never heard anything afterwards.

I told Thom about this and he assured me that deer do not eat grass, so it was definitely something else out there. He then told me that the flowers growing all around my tent had seed pods that were edible and we walked over to check it out. That had to have been what was being plucked. I never noticed a large wad of anything that had been pulled out of the ground or any damage to any of the flowers, so they must have been harvesting very carefully so as not to harm the plant itself. I was very impressed with this since I know people who aren't that careful at harvesting and will just rip a plant to shreds while collecting the fruits the plant has to offer.

The next evening, I again heard the sound of heavy footfalls around my tent. This time they were coming in from both sides of the tent and I felt a bit intimidated by them. They were heavier than what I had heard previously and the thought that went through my mind was that the adults had just entered camp. I then heard a "tink" sound... like a rock hitting a rock. It was a small sound and came from behind and to the left of my tent. As soon as I heard it I knew exactly where it came from. Under the tree directly behind my tent I had spotted an area with rocks laying in the dirt. It had caught my eye because the rocks were chipped in various spots and had rock dust laying on a few of them and it looked like they had been hit together in the past.

A few seconds later I heard it again... tink. I just smiled and waited to see if it would come again. And it did… tink. Each sound was exactly the same and came from the same spot. I remember wondering what they could be doing back there. Then I heard footsteps coming from in front of my tent, continuing on along the left side of it, and then heading back towards where the rock sounds were coming from. I heard rustling sounds from the tree back there for a while, but never heard any vocals of any kind. I then heard a branch snap come from way back in the treeline. At that sound, both sets of footsteps walked away from the back of my tent, went up the dry creek bed and off into the trees.

I checked the area for signs of something being in the area when I arose the next morning, but didn't see any footprints of any kind. I thought that was really odd since there are areas of soft dirt here where the vegetation thins out. I thought for sure there would be a few tracks left, or perhaps I would see a hand print at least. But there was nothing there at all that I could make out.

One night before putting the fire out, I stood in front of it thinking about how I was in for a long night. The sasquatch had been spotted on the hill across from our camp kitchen during the day. I knew that if they were getting brave enough to be seen during the day, they were definitely not afraid to come on into camp at night and do whatever they wanted to do. I had no regrets about pitching my tent so far away, but it sure was a long walk to get there after I put the fire out each evening.

I didn't light any lights when I got into my tent that night. I was pretty tired and the tent was lit up enough that I could actually see well enough to get ready for bed and climb into my bag. While I was laying there I once again heard the sound of something large walk up to my tent, and this time they came all the way up to the front corner of it. I then heard what sounded to me like branches rubbing across the top of the tent, or perhaps pine needles scattering across the top of it. I looked up to see a shadow in the shape of a very large hand and arm making a slow circle across the top of the tent!

The sounds I had heard had to be the hair sliding across the nylon rain cover that was covering the mesh ceiling. It made two passes across the top and then ran its hands across both sides of the tent. It hadn't moved while touching both sides of the tent and I was left to wonder just how large this thing really was.

My tent isn't large by any means, just an eight by eight tent, but my mind was blown away by the fact that it could reach around both sides of it at once like that. That pretty much makes my tent just the right size to be picked up and run away with. This thought certainly gave me pause that night. Here I was wrapped up like a sasquatch burrito laying in a take home doggy bag... and here was something large enough to carry it! As I was laying there thinking about this, I heard a short whistle come from the woods behind me and the one standing over my tent walked away in that direction.

Another night out there I was just sitting and reading and something walked straight up to my tent and poked the side of the tent opposite me. It wasn't an aggressive move, just poked the tent slightly. It surprised me and I stifled a squeal and tried to continue reading as though nothing was happening. I started to read the same sentence again, and it pushed on the tent again. This one was a little bit harder. I once again began to read the same sentence that I had been reading before.

I heard it walk around the back of the tent and give it a little poke. I kept telling myself to just stay calm and kept rereading the same sentence while trying to get my mind off of what was going on around me. Then it walked around the corner of the tent and it pushed in on the side of the tent where I was sitting. This scared me a bit and I involuntarily moved over closer to the middle of the tent and found myself off the sleeping bag entirely.

I once again began to read the same sentence over, and it came again. This time it pushed really hard on the tent and I saw the side begin to buckle. I found myself getting a bit concerned that it would break my little tent just trying to get a reaction from me. Not knowing which type of reaction to give it, I tried to continue reading my book. I wasn't reading out loud and I knew it was just a coincidence that every time I began reading, it began poking.

It then walked around to the front of the tent and I know my eyes began to bulge out of my head. I wondered if it knew how to unzip the tent and if it had planned on coming in with me. This thought concerned me because I didn't know this clan and I didn't know what their intentions really were. It just pushed the tent in again and then stood there.

By this time, I had read the same sentence in my book about twenty times, and I let out a slight laugh at how foolish I was being. This seemed to spur it on and it pushed super hard on the front of the tent and I noticed the left side sag a bit. I opened my mouth to tell it to quit pushing so hard when I heard a soft whistle come from in the woods. The one by my tent wandered off in that direction and I was left to wonder which of the clan it was that kept whistling for the others. I figured it had to be the matriarch or someone just as important because everyone listened to it instantly.

The next morning around the fire Arla remarked on how tired I looked. I told her that I was exhausted and hadn't been getting much sleep with the sasquatch coming in every night. She told me to just set my boundaries with them. She said that I had to let them know that I needed to get some sleep, and to ask them to just leave me alone for a night or so. I had no idea that I could just tell them something and they would listen, and I told her that. She advised me that we do have some control over the situation when it's happening and that they would listen to us as long as we treated them with respect. I thanked her for the information and then mulled it over for the day. I had never even thought about this sort of thing and I had no idea how well it would actually work, but she knew more about the sasquatch than I did and I decided to try it that night.

I was thinking about this as I put the fire out that night and decided that it wouldn't hurt to give it a try. If it worked then I could get a good night's sleep and have more energy the next day. I walked up to the front of my tent and stood there. I began speaking out loud, not really knowing if there was even anyone there to hear what I had to say. I stood in front of my tent and said, "I love you guys, and I love having you come around, but I am so very tired and I really need to get some sleep tonight. Would you mind leaving me alone, just for tonight? I know someone else in camp who would love to meet you and have an interaction with you!" At this point I raised my flashlight and pointed it to another campsite that contained a very nice couple, Russ and Nancy, who had come up and camped with us. I said, "Just go on over and play with them tonight. I know Nancy would love to have you in camp. Just go over and say hello and introduce yourselves", and I shined my light on their camp once again.

I said, "Well, goodnight and I'll see you tomorrow night. Okay?" I unzipped my tent and crawled into it. Since I already had my thermals on under my clothes, it took me under two minutes to be undressed and in my mummy bag. As soon as I got into it and got comfortable I heard three different sets of footfalls walk right past my tent and across the meadow to the left directly towards Russ and Nancy's campsite that I had just lit up with my light! I was both stunned, and amazed! I just could not believe that it had worked, and I just could not believe that they had actually been out there listening and watching and waiting for me to climb into my tent! I really had no idea that they were out there.

When I approached the kitchen fire the next morning I heard everyone talking about how Nancy had seen a sasquatch standing on the other side of her truck the night before. She was going on about how exciting, yet scary, it had been for her, and how her husband, Russ, had never looked out of the hammock he was sleeping in or got out to take a look. He then told us about how he had felt something touch the underside of his hammock, and he had no desire to come out and say hello. He had set the hammock up at the edge of a drop off above the creek and something had come up the backside of him and touched him. They were both so excited and I was so happy for them.

I didn't want to attract attention to myself and was just standing there smiling when Arla, who given me the advice told them, "It was Kathi's fault. She sent them over there!" Talk about being busted! At least it wasn't for something that would make them angry and we all had a good laugh over it. I for one was still amazed that it had worked, and here was all the confirmation that I would ever need. You could talk to the sasquatch and have them respect you enough to honor your wishes… if the request was given with respect in the first place. What a concept!

The sasquatch did go to the other campsite and they did stand there and introduce themselves. Later on during our camp trip this Nancy and Russ ended up with a very clear picture of one of the sasquatch in the brush by their campsite. I was so happy for them because they are really good people and they deserved this experience.

The next morning we drove into Willow Creek. I was very excited to go and we had a great day. We ate breakfast at the Bigfoot Restaurant which is located across the street from a really weird mural that is painted on the side of a building there. This mural depicts the sasquatch helping man do different chores, like cut wood, build a building and plant trees. No wonder they stay away from us, we work them death!

We went to the Willow Creek-China Flat Museum also. The bigfoot exhibit contains a lot of really interesting pictures and old newspaper articles, along with a lot of sasquatch collectibles. One of the things that I really enjoyed viewing there was the foot print casts that they have on display. It's easy to see how different each foot really is when you see them all displayed in one place like that.

We then drove to a campground located on the Trinity River to meet with another group of campers. It turned out that they were all friends from a Facebook group that I wasn't familiar with. They were all so nice that I joined the group as soon as I got home. I walked down to the river to get cooled off a bit and found a rock that is shaped just like a foot. It was a bit heavy, but it was so unique that I had to have it. I lugged it back up the steep hill to the campground and stashed it in Thom's car to take home with me.

Figure 59 Mural on building in Willow Creek

A few of our fellow campers decided to stay in Willow Creek, but three of the other group wanted to come up to our camp for a night to get in on the action we were having there. They decided to drive up the following morning, so we left and made our way back to our camping area.

We didn't get back to our camp until nearly sunset and it was so nice to pull in and see a fire already lit at the camp kitchen for us. Russ and Nancy had come back earlier than we had and took the time to get it all ready for us and we were ever so grateful for them. There were only five of us in camp now that the others had stayed in Willow Creek and we sat around the fire talking.

Eventually, everyone wondered off to bed and Thom and I sat at the fire listening to the sound of screams (wailing) coming down the mountain behind us. It was the first time I had heard that type of vocalization and I could certainly see how scary that sound could be if you were alone in the forest. They were far enough away that I could just catch the sound of it, so I assumed that they

Figure 60 Car at camp kitchen, road and berm

had business elsewhere and we would be alone in camp that night. I didn't stay out by the fire for long after Thom said goodnight as I was pretty tired myself.

There was no moon present in the sky that night and it became very dark after the sun went down. Instead of just cutting across the grass to get to my tent like I had previously, I elected to walk out onto the dirt road that cut through the middle of the meadow and follow that to my area. I didn't want to chance turning an ankle tripping over the many rocks that jutted out of the grass.

There are trees growing along one side of this road which end abruptly where a dry creek bed makes a bend and comes down off the mountain. It is very deep with large rocks at the bottom of it. It's impossible to see into the side of the meadow where my tent was from the road until the trees end and you are instantly dumped into it. I will include a few picture's here which were taken from inside my tent to show the lay of the land that night so that you can more easily follow along with what happened next.

I had turned on my flashlight as I was walking down the dirt road, but kept it hanging down towards the ground to light up my feet and the road in front of me. It wasn't my wand light, which would light up a football field, it was a maglight with only a small circle of light. I hadn't heard any sounds at all since Thom and I had heard the wailing earlier in the evening. There were no footfalls or whistles, so I assumed that we were all alone out there as far as the sasquatch were concerned. Just goes to show how quiet and elusive this species can be when they want to be! I'm still not sure how they can be so loud and you can actually feel their footsteps hitting the ground when they want you to, and be so quiet and stealthy that you never even know they are there when they don't want you to.

I walked out of camp with my flashlight pointing down at the ground and walked up the dirt road along the treeline. At one point I remember stopping and just standing there and looking up at the stars. They appeared to be so close you could just reach up and touch them. I remember saying how beautiful it was up there, and how much I was going to miss it when I went home in a few days. I didn't want to think about leaving just then, so I dismissed that thought and continued walking up the dirt road to my part of the meadow.

When I arrived at the little tree, I lifted up my mag light and shone it across the field. I moved the light slowly from one end of this area to the other twice. Once at ground level and then back over it again a bit higher up. I was actually looking for eye shine from a bear or a deer as I wasn't expecting any sasquatch to be out there. The first time I swept the meadow, I saw my tent reflecting the light back out at me and there was nothing in front of it or near it. There was also nothing on the ground in the meadow. The second time I swept the meadow there was my tent reflecting back at me with nothing at all in front of it, or near it at approximately five to six feet off the ground. I didn't shine the light up any higher or up into the tree behind my tent.

As my light hit my tent the second time, I remember saying out loud, "Oh, there's my tent." I wanted to talk out loud in case there were bear in the tree line. I knew in my heart that there was nothing out in the middle of the meadow because I was able to see every flower and every weed growing on the ground there. I was taking a close look for bear or deer or any eye shine that I may see before I walked out there, and I thought I had scrutinized the area really well.

I was still standing on the dirt road at this point, and hadn't completely walked out into the open area beyond the last little tree. That would be the new growth that you can see to the right of the picture above. Once I was convinced that there was nothing eating out in the meadow, I walked past the two new growth trees and was now standing to the left of them directly in front of my tent. (That would be where the gray rock pops up out of the ground to the left in the picture.)

I stopped at the top of the runoff trench to quickly scan the bottom with my light so that I could take stock on where all the large boulders were located, and then continued sweeping my light up the other side of this trench to calculate exactly what path I was going to take to get through here, all in one continuous motion. As this wasn't a wide trench, this took perhaps a fraction of a second to maneuver. I still had not heard a single sound out of the night at this point. It would have taken me approximately eight steps to get from there to my tent.

As soon as my light came up out of the trench and hit the front of my tent, I saw a sasquatch standing there. And in the wink of an eye it turned and sort of hopped over the corner of my tent and ran towards the tree that sat behind it. All I really remember seeing were two tree trunks turn (or swivel) and one tree went around the left corner of my tent, and the other one followed it and then they took three steps and stopped behind the tree. I not only saw these steps, but I felt them hit the ground. Hard to actually describe such a heavy, flat, thudding sound to people, but I will never forget exactly what it sounded like.

I don't remember looking into its face, but I know I did. I just can't bring it to the front of my mind at all. The only thing my mind wants to remember seeing is its legs moving around the corner of the tent. My mind wanted to identify them as "tree trunks" because the legs were easily the size of a medium tree, and about the color of Redwood. But then my eyes caught sight of the ankles…and we all know that trees don't have ankles! I then saw the hair at the bottom of the calf float up slightly as it raised its leg and then put it back down again. It was the most awesome sight I had seen to date. This baby was moving swiftly too!

I say that I saw two tree trunks because that is exactly what my mind wanted to describe them as. Thick, redwood colored, tree trunks. Except for the fact that these tree trunks had ankles! The legs tapered down and I saw definitely saw ankles as it lifted its legs to hop over the side of the tent. And these tree trunks were covered in hair that swept up above the ankle as it ran away... Not like it was blowing in the wind, but just slightly puffed up away from the leg as it turned and jumped over the edge of the tent. I think that the reason I remember this part so vividly was because my mind worked so hard trying to identify them as tree trunks, and not legs, and it was having a really hard time doing so. I would love to be hypnotized and see if I can remember what his face looked like.

It all happened so fast that I barely had time to log it all, but the one thing that my mind did grab hold of was the fact that it had only taken three steps and then had planted itself right behind that tree. And that tree was only sixteen steps away from me. I know this because my tent is eight by eight and the tree sat directly behind it! I stood and listened for it to continue running off in the opposite direction, but he stayed right there watching me.

I know I must have stood stunned for a few minutes just gathering my thoughts and waiting for it to run off. When I finally did come back to my senses my first thought was that if it wasn't leaving, then I'd better apologize! And that's exactly what I did. I said, "Oh, I am so sorry Brother or Sister! I didn't see you standing there, or I would have never shone the light at you. I'm sorry to scare you, but you know you just scared the (heck) out of me too! I think I just lost a year or so off my life, but I'm so sorry if I scared you! If you want to stay here in the meadow and eat, or whatever you were doing, it's just fine by me. I'll just go back over to the fire and let you do what you came here to do! And I am so sorry if I scared you, but you have to admit that we just scared each other."

I chuckled then to let it know that I was still in a good mood, and maybe that would lessen any anger he/she may have been feeling.

At that point I had intended to just swivel around and head back to the camp kitchen area and build up the fire again and let them have the meadow. I had absolutely no problem with that thought at all, but as I went to swivel around I noticed that I couldn't do it. I tried to get my feet to back me up so I could get out of there, but they wouldn't move either. I started to get a bit frustrated at that point and I again started blabbering out loud and said, "I really am sorry Brother or Sister and I didn't mean to come up on you like that. I want good will and I mean you no harm. I'm going to go back over to the fire and leave you alone in the meadow now."

Once again I found that I could not get my body to move on command and I noticed that my head was now looking at the ground and my hand was pointing the light down toward the ground also. I have to admit I was starting to get plenty irritated at this point but even I'm not hard headed enough to snap at something this potentially dangerous, so I did the only other thing I could think of and that was to call Thom. I yelled out, "Thom! Could you come here please? I need some help out here! Right now!" I knew that he was already in bed and I was quite happy to hear his voice ring out in the night telling me that he was on his way.

I was still apologizing profusely, and stuck in place, when Thom arrived. When he came up behind me I heard him say something, but all I caught was something about how big this one was. Well, I knew that, but I sure appreciated someone else being able to see it and substantiate my claim. I just hoped that Thom could talk to this one and see if he was okay, or if he meant us any harm. Thom began talking to it and I asked him if it was male or female, and he replied that it was a male, so I started saying, "I'm so sorry Brother, it wasn't fair of me to walk up on you like that. Don't be mad at me or take it out on me, okay? Sorry that we scared each other."

Yes, I was babbling at this point, and no, I didn't notice how much at the time. It just felt good to be able to address it specifically instead of saying Brother/Sister all the time, and I wanted to make it more personal for him. At one point Thom began to chuckle and I asked him what was going on? He replied, "You're both apologizing to each other at the same time and neither one of you is listening to the other and there is a complete lack of communication here."

That made me stop and think about what was going on, and I sure wished that I had learned to open my mind up more so that I could communicate with this guy. It made me sad that I couldn't hear him. Thom told him that I was just going to walk over to my tent and get my stuff and I would sleep with him in his tent that night. He gave me a slight nudge while leaning in towards me and whispered for me to walk over and get my stuff from my tent. I rocked forward a bit when he nudged me, and replied that the reason I called him was that I couldn't will my body to move for me. I was stuck!

He then spoke to the sasquatch again and told him that he would take responsibility for me and would he please release his grip on me so that I could get my stuff out of the tent. It was the strangest thing that had ever happened to me when I felt my body go limp. I had to correct myself and take hold of my body again so it didn't just drop to the ground. I was buzzing with electricity. My entire body was tingling like I had just been shocked.

I handed Thom my walking stick and my flashlight and walked to my tent. I unzipped it and knelt in front of the door and just pulled my bag from the tent and zipped it back up. I then walked back to Thom and we continued back across the meadow to his tent. The sasquatch never moved the entire time and I don't remember ever hearing him make a sound either.

When I got myself situated in Thom's tent and was lying there thinking about what had just happened, I heard Thom ask me if I was okay. I told him that my body was still tingling like I had just received electric shock therapy, but other than that I felt fine. He then said out loud, "Yes, she is doing okay now."

I asked him what that was all about and he told me that the sasquatch was checking in and asking if I was okay. I thought that was about the sweetest thing ever! To actually care about my feelings! I know humans who don't care that much about me, and it touched me deeply.

I asked Thom if he knew his name, and he told me what it was. I sure wished then that I could speak to him and tell him how much it touched me that he was worried about my well-being, when all I could think of was his well-being and had been hoping that I hadn't offended him in any way. My fingers and toes were the only things tingling on me at that point and I had to keep rubbing them trying to make the feeling go away.

After Thom fell asleep, I could hear numerous footfalls around the tent and out around the camp kitchen area. I then heard the hummingbird/bee sound moving around the inside of Thom's tent. This time I got a smile on my face and I said very quietly, "Oh sweet! That sound was from you guys, huh? Thanks for checking up on me, I'm doing fine. I hope you are too!" I then heard a brief communication happening outside and I listened to them converse for a few minutes.

Fortunately, Thom's tent is the size of a small cabin, so there was plenty of room for him and me along with Jackie to sleep quite comfortably together. Unfortunately, I had forgotten to grab the pad underneath my bag, so I felt every rock and every stick on the ground underneath me that night and never really did get any sleep. As soon as I saw the sun light up the sky, I grabbed my bag and headed for my tent so that I could get a few hours of sleep before everyone else woke up.

All too soon I heard voices approach my tent and stand next to it talking about what had happened the night before. As soon as I opened my eyes, I knew it wasn't going to be my best day ever. I think I got three hours sleep total and awoke very bleary eyed. Just trying to get my shoes on was the hardest work I had ever done, and I'm not proud of how I must have looked falling out of my tent. Everyone was standing there looking at me and then started asking me questions.

I think I just grunted because my tongue was stuck to the roof of my mouth at that point, and the only thought I could grab hold of was…coffee! So I just ambled off towards the camp kitchen to grab some and left them all standing there talking. I apologize to everyone who was there that morning. I apologize for my attitude. I apologize for the grunt and mostly, I apologize for the way I must have looked as I fell out of the tent. I'm betting that was the scariest part of everyone's encounters on that trip!

As I sat drinking my coffee, one thought was pestering me to no end. At some point in the night it had occurred to me that the sasquatch had to have known that I was coming in to that meadow! It couldn't have been a surprise to it that I was there... Not like it surprised me to find him there. I had shined my light all over that meadow from one end to the other and had clearly seen my tent twice. The only thing I could think of was that perhaps it had been hunkered down next to my tent on the blindside. But if it was on that side of the tent, it would have to have been down on all fours for me not to have seen its head poking up above the tent. If that was the case, why didn't it scoot off behind the tree before I stepped out into the open? It could have easily run along the side of the tent and under the tree from where it would have been located, and I would have never seen it. And why was it standing in front of the tent when there certainly hadn't been anything there three seconds earlier?

These thoughts have plagued me ever since I had this experience. I still don't understand the entire occurrence. Did it want me to see it? And if so, why did it run off so quickly? If it hadn't wanted me to see it, why hadn't it hidden itself in the time it took me to get to my tent area? I had the light on the entire time I was walking and I was quite visible from the moment I walked away from the fire. I even stood on the road admiring the stars for a while. It had plenty of warning that I was coming because I spoke out loud on the road, and I spoke out loud in front of the

Figure 61 Loaded Car

meadow.

My thoughts keep going back to the fact that it wasn't standing there when I shone my light on my tent twice, and it couldn't have taken me more than a few seconds to go from that to shining my light through the trench. That's basically the one thought that still nags me.

All too soon, it was time to pack up the car for the long ride home. This time we had three people packed into Thom's car along with all our gear. We ended up sitting on some of it and had some of it on our laps too! To say that it was a pretty uncomfortable ride would be making light of it, but we managed quite well with only a small amount of complaining.

On the way home we stopped and camped for the night in a small pullout area Thom found in the mountains next to a river. I mention this part, even though no sasquatch were in evidence, because for me it was another one of the highlight's of our trip. I woke around four am and needed to get up and stretch, so I got dressed and walked out onto the road. I noticed a bridge up ahead a short ways and walked down to look at the river. No cars ever passed by, and I was completely alone.

The sun was just coming up over the top of the mountains and the swallows were all flying around catching breakfast. I stood watching them for quite some time just enjoying the beauty of the river and the surrounding countryside. With the complete absence of all people and traffic, I was reminded of the movie "The Omega Man." It was so cool to be able to stand right in the middle of a major roadway in California and not be run down, and I felt like the last woman on earth.

All too soon it was time to wedge ourselves back into the car and continue homeward. We came upon an elk viewing area and stopped to see what was out there. All I could see were the cows, but Thom showed me how blind I can really be. He handed me his binoculars and told me just where to point them to find each of the bulls that were there also. Sure enough, there they were laying down in the high grass. I love spending time with this man, because I learn something new each and every time I'm with him.

Thom was even kind enough to stop at the Redwoods on the way home so we could see the beauty of that area too, so even though the drive was a bit cramped, it was still a trip I wouldn't have missed for the world.

My journey up until this point has been a great adventure, and it's not over yet. I will spend time in the forest with these beings until I am no longer physically able to get there. After that, I will relive the wonderful things I have experienced through my memories. I hope that doesn't happen for many years to come because I can still hear the forest calling me.

MAY THE FOREST BE WITH YOU!

Epilogue

Now that I have been doing this for some years now, I wanted to add this chapter to explain how I now feel about some of the things I said in the earlier chapters, since I now realize how wrong I was about a few things.

I no longer feel that it is inappropriate to bring children camping in an area of known sasquatch activity. In fact, my friend, and research partner, Cristy now has two children of her own and we wouldn't dream of going camping without them. However, I do think that if you bring your children into the woods you should remain very aware of them at all times while they are out there with you. I have read no current reports about the sasquatch abducting women or children and have, in fact, heard many reports of sasquatch youngsters playing with our children and having a great time. I have even heard of cases where the sasquatch will lend aid to wounded hikers and fishermen. The reason I advise caution however, is the many recorded cases of children disappearing without a trace right under their parents noses while out enjoying the wilderness. It isn't just the sasquatch that is living out in the woods, and a bear or mountain lion would be much more likely to run away with a child.

I do still feel that our pets are best left at home however, and I have seen many pictures of what can happen to dogs if they disrupt the sasquatch in any way. It is not pretty and I'm very sure these dogs suffered greatly. If you love Fido, keep him at home while you are out spending time with the sasquatch. I do know of a few researchers who take their dogs out into the field with them and nothing bad has happened to them as of yet. Perhaps the answer lies in the dog itself and what sort of attitude it may have towards other beings, and what sort of mood the sasquatch is in when they come together.

I have also changed my mind about how deep into the woods one must go to have an interaction with the sasquatch. When I first found the tracks in our county park, I wouldn't allow myself to believe that they could be so close to civilization. I now know better. Not only can they live so close to civilization, they do live so close to civilization. They will even follow you home if they so desire to do so! Many people have described to me how objects will appear, or disappear, at their homes and they will find tracks in association with the find. I even know a lot of people who have the sasquatch come into their pastures and leave glyphs for them to find. And if the homeowners leave a glyph of their own, the sasquatch will change the design into something else! Which also shows me that the sasquatch have a playful side.

I also know that not all of them are the monsters that they are portrayed to be by the media, which is what I thought they were when I first started having contact with them. I know that some of them don't like humans, and would rather have nothing to do with us. These may go out of their way to abuse us petty humans, but I'm willing to bet that they are few in numbers. I also know that the sasquatch will try to intimidate us into leaving an area if they feel there is a reason for us not to be there. They may throw rocks or sticks at us, or growl and circle us. I have never had that experience personally, but have heard from others that this has happened to them. Since they all lived to tell the tale, I can only assume that the sasquatch had a reason for this and it was best that the humans left the area immediately.

I also want to explain the reason that I destroy each and every print I find now. Since I was gifted with the one perfect print to hang on my wall, I'm no longer interested in casting them. I don't want to leave them lying there and alert anyone else to their presence either. I don't want anyone to come along and track them back to their home base, and I sure don't want any of them to be shot at.

I'm happy with the casting that I have, and it is enough for me. When I'm out in the forest now, I just have fun. I know they are watching me and I never want to do anything that may make it look like I'm hunting them. I don't go out there with the mindset that I have to find proof to take to others. I go out there with the mindset that I want to enjoy my time with them in the beauty of the forest.

Since I have learned to leave all the electronics at home, they have come closer to me than they ever did while I was using the recorders, the video cameras and the night vision. I don't know how they know when we are operating these things, I just know that somehow they do and it will put a halt to them coming around.

Oh, and one last thing. While out in the forest, please remember that this is someone else's home! I have spent so much time in the wilderness having to pick up other people's trash before I could settle in for a nice stay. Nothing is more heart breaking than to look around a beautiful forest glade and spot paper garbage and cigarette butts and car batteries and dirty diapers dotting the landscape. Please carry out ALL of your garbage. Why would you want to litter such beautiful forests anyway? Do you go to the wilderness to see other people's garbage? I would rather see the beauty that the creator has made.

And if you have to use the restroom, please dig a hole and bury your paper and excrement. That's not only an eyesore, but it poses a health risk to everyone who comes in after you. I'm not asking you to dig an entire latrine, but at least scuff up the top layer of soil and cover up what you've created. Thank you.

Sasquatch doesn't pollute our homes…Let's not pollute theirs!

www.ingramcontent.com/pod-product-compliance
Lightning Source LLC
Chambersburg PA
CBHW051855170526
45168CB00001B/114

9781512083590